Research for Development

The series Research for Development serves as a vehicle for the presentation and dissemination of complex research and multidisciplinary projects. The published work is dedicated to fostering a high degree of innovation and to the sophisticated demonstration of new techniques or methods.

The aim of the Research for Development series is to promote well-balanced sustainable growth. This might take the form of measurable social and economic outcomes, in addition to environmental benefits, or improved efficiency in the use of resources; it might also involve an original mix of intervention schemes.

Research for Development focuses on the following topics and disciplines:

Urban regeneration and infrastructure, Info-mobility, transport, and logistics, Environment and the land, Cultural heritage and landscape, Energy, Innovation in processes and technologies, Applications of chemistry, materials, and nanotechnologies, Material science and biotechnology solutions, Physics results and related applications and aerospace, Ongoing training and continuing education.

Fondazione Politecnico di Milano collaborates as a special co-partner in this series by suggesting themes and evaluating proposals for new volumes. Research for Development addresses researchers, advanced graduate students, and policy and decision-makers around the world in government, industry, and civil society.

THE SERIES IS INDEXED IN SCOPUS

More information about this series at http://www.springer.com/series/13084

Niccolò Aste · Stefano Della Torre ·
Cinzia Talamo · Rajendra Singh Adhikari ·
Corinna Rossi
Editors

Innovative Models for Sustainable Development in Emerging African Countries

Editors
Niccolò Aste
Architecture, Built Environment and Construction Engineering—ABC Department
Politecnico di Milano
Milan, Italy

Stefano Della Torre
Architecture, Built Environment and Construction Engineering—ABC Department
Politecnico di Milano
Milan, Italy

Cinzia Talamo
Architecture, Built Environment and Construction Engineering—ABC Department
Politecnico di Milano
Milan, Italy

Rajendra Singh Adhikari
Architecture, Built Environment and Construction Engineering—ABC Department
Politecnico di Milano
Milan, Italy

Corinna Rossi
Architecture, Built Environment and Construction Engineering—ABC Department
Politecnico di Milano
Milan, Italy

ISSN 2198-7300 ISSN 2198-7319 (electronic)
Research for Development
ISBN 978-3-030-33322-5 ISBN 978-3-030-33323-2 (eBook)
https://doi.org/10.1007/978-3-030-33323-2

This Springer imprint is published by the registered company Springer Nature Switzerland AG
The registered company address is: Gewerbestrasse 11, 6330 Cham, Switzerland

Preface

This book belongs to a series, which aims at emphasizing the impact of the multidisciplinary approach practised by ABC Department scientists to face timely challenges in the industry of the built environment. Following the concept that innovation happens as different researches stimulate each other, skills and integrated disciplines are brought together within the department, generating a diversity of theoretical and applied studies.

Therefore, the books present a structured vision of the many possible approaches—within the field of architecture and civil engineering—to the development of researches dealing with the processes of planning, design, construction, management and transformation of the built environment. Each book contains a selection of essays reporting researches and projects, developed during the last six years within the ABC Department (Architecture, Built environment and Construction engineering) of Politecnico di Milano, concerning a cutting-edge field in the international scenario of the construction sector.

Undoubtedly, the African continent will see the most interesting trends in the near future for the construction sector, as well as the most serious risks in terms of sustainability of the development models. These countries face two parallel challenges: fighting the lack of resources and channelling their development along a sustainable path. In both cases, innovative methods and technologies can offer a significant contribution: affordable housing set within the social context should develop in parallel with a wise exploitation of the energetic resources; the sustainability of the entire system partly depends on how waste is handled and how to set up a virtuous recycling system; emergency situations must be addressed rapidly and efficiently, and the introduction of low-cost technologies may allow to turn study and preservation of the cultural heritage into an opportunity for development, without subtracting resources from humanitarian assistance. In general, connecting past and present will help to shape the future of countries, where carefully chosen innovative instruments can really make the difference and can allow giant leaps towards a sustainable social, cultural and environmental balance.

The book presents a selection of innovative projects carried out in African countries, aiming at tackling two main areas: offering practical solutions to specific necessities and experimenting cost-effective methods and technologies, which could be easily applied in order to achieve high-quality results. The papers have been chosen on the basis of their capability to describe the outputs and the potentialities of carried-out researches, giving a report on experiences rooted in the reality and at the same time introducing the perspectives for the future.

Stefano Della Torre
Head of the Department Architecture
Built Environment and Construction Engineering
Politecnico di Milano
Milan, Italy
e-mail: stefano.dellatorre@polimi.it

Introduction to Volume

Estimating population, urban and economic growth for Asian and African countries (United Nations 2014) highlights the need to draw up new design approaches that promote the development of original architectural languages appropriate to the local identity and to the new development models. These models support the consolidation and growth of local cultures and economies and at the same time aim at reducing energy consumption, minimizing environmental pollution, increasing the use of renewable sources and effectively responding to water demand.

Therefore, a change of paradigm is required, together with a novel approach to urban (and peri-urban and rural) planning and usage of territories. In this perspective, a holistic view should influence the entire built environment, i.e. the configuration of goods, the structure and use of land and the way in which basic services—such as energy, water, food and waste treatment—are handled. It is about working to move away from the current model of linear urban metabolism—based on the 'take-make-dispose' approach—to a circular one, where the consumption of resources and the waste production are minimized.

Combining these changes into practice requires working in two main directions. It means, first of all, focusing on the relationship between the architectural, urban and physical aspects of new developments, climate and energy demand. Secondly, it requires identifying and integrating the necessary strategies and infrastructures to close off the energy–water–food–waste circle, searching for a high level of efficiency and self-sufficiency.

The present book revolves around these issues and describes the contribution of the ABC Department of Politecnico di Milano regarding the multidisciplinary research and development (R&D) activities, carried (and being carried) out on these multifaceted issues. It presents a selection of innovative projects carried out in African countries, aiming at tackling two main areas: offering practical solutions to specific necessities and experimenting cost-effective methods and technologies that can be easily applied in order to achieve high-quality results.

Emerging countries face two parallel challenges: channelling their development along a sustainable path and fighting the lack of resources. In both cases, innovative methods and technologies can offer significant contributions, presented in this book:

affordable housing set within the social context should develop in parallel with a sustainable exploitation of the energetic resources (Part I); the sustainability of the entire system partly depends on how waste is handled and how to install a virtuous recycling system (Part III); emergency situations must be addressed rapidly and efficiently, and the introduction of low-cost technologies may allow to carry out study and preservation of the cultural heritage without subtracting resources from humanitarian assistance (Parts II and IV). In general, connecting past and present (Part V) will help to shape the future of countries where carefully chosen innovative instruments can really make the difference and can allow significant leaps towards a sustainable social and environmental balance.

Cinzia Talamo
Niccolò Aste
Corinna Rossi
Rajendra Singh Adhikari

Contents

About the Editors

Niccolò Aste Full professor of Building Physics and Building Energy Systems at the ABC Department of the Politecnico di Milano. Graduated *cum laude* in Architecture and earned a Doctorate (Ph.D.) in Technological Innovation and Architectural Design at Politecnico di Milano. Starting from 1995, he has been actively engaged in research activities related to energy efficiency of the built environment, and to the exploitation of renewable energy sources and their use in building sector, both at urban and at territorial level. He participated in several national and international research programs on the correlated issues of sustainable architecture, energy efficiency and renewable energy sources. Among major clients of his research consultancy are: United Nations (UN-Habitat), Italian Ministry of the Environment, Lombardy Region, FIAT, GSE, Whirlpool, ENI, ENEL. He is the author and co-author of several scientific handbooks and more than 100 publications in various international and national journals, and regularly participates to conferences and seminars.

Stefano Della Torre Graduated in Civil Engineering and in Architecture, he is a Full Professor in Restoration at the Politecnico di Milano, and the Director of the ABC Department - Architecture, Built environment and Construction engineering. He is the author of more than 250 scientific publications, and serves as advisor for: CARIPLO Foundation (Cultural Districts), Province of Como and Lombardy Region (policies of programmed conservation of historical-architectural heritage). He is the President of BuildingSMART Italia, the national branch of BuildingSMART International.

Cinzia Talamo Full professor of Architectural Technology at the ABC Department of Politecnico di Milano, she hearned there a Ph.D. in Technical innovation and design in architecture and a Master Degree (five years program) in Architecture, awarded with honors. She is the President of the Sub Committee U/CT025/SC03 Maintenance of Real Estate and Facility of UNI (Ente Italiano di Unificazione, the Italian institution dedicated to standardization). She is the Coordinator of the Scientific Committee of the ABC Department. Her main and long-term interests focus on planned maintenance, as well as building and urban facility management. A recent area of interest concerns cross-sector waste recycling in the perspective of industrial symbiosis and circular economy.

Rajendra Singh Adhikari Associate Professor of Building Physics and Building Energy Systems at the ABC Department of the Politecnico di Milano. He graduated in Physics and earned a Ph.D. in Advanced Technologies for Solar Energy Utilization and Master in Energy and Environmental Economics and Management. His research work mainly focuses on renewable energy, energy efficient buildings and energy conservation. He participated to various national and international research projects, including IEA activities (SHC-Task 25, Task 38 and PVPS-Task 10). He is the author and co-author of over 100 publications in various international and national journals, and regularly attends conferences and seminars. He was member of the technical committees of various international conferences (PLEA 2011, PLEA 2012, PLEA 2015, IEECB'14, IEECB'16), and is a Life member of ICTP-India Chapter (International Centre for Theoretical Physics, Trieste, Italy).

Corinna Rossi Architect (University Federico II Napoli) and Egyptologist (M.Phil, Ph.D. Cambridge University), is Associate Professor in Egyptology, Director of the Italian Archaeological Mission to Umm al-Dabadib (Egypt), Principal Investigator of the project L.I.F.E. (Living in a Fringe Environment) funded by the ERC Consolidator Grant no. 681673. She is the author of several publications on the relationship between architecture and mathematics in ancient Egypt, and on the relationship between ancient and modern measurements. Since 2001 she has been working on the first scientific exploration of the Kharga Oasis, in Egypt's Western Desert, where she discovered and documented several archaeological sites.

Housing Models

Introduction

Cinzia Talamo, Niccolò Aste, Corinna Rossi, Rajendra Singh Adhikari

Over the years, the concept and practices of a sustainable built environment keep evolving. One aspect that is being increasingly acknowledged is that theory and practice should be expanded beyond individual buildings and spread to the scale of the entire community. Sustainable community design and practices request complex interactions between cultural, social, economic and environmental factors in order to meet sustainable development objectives.

Cities of developing countries in tropical regions, particularly in Africa, are characterized by a staggering urban population growth, a rocketing increase in energy request and consumption, and by equally fast-growing environmental issues. In this scenario, active strategies and building energy optimizations are strongly recommended and should be accompanied by the elaboration of new and innovative housing models.

The following section describes objectives and results of some research projects aiming at offering practical and sustainable solutions. The main issue, hovering above the discussion, is the necessity to integrate tradition and progress, both at the social and at the technical level, in order to design a sustainable path towards an efficient development.

BECOMe—Business ECOsystem Design for Sustainable Settlements in Mogadishu

Oscar E. Bellini, Andrea Campioli, Claudio Del Pero, Cinzia Talamo, Nazly Atta and Anna Dalla Valle

Abstract The paper introduces the "BECOMe" project, winner of the PoliSocial Award 2018. BECOMe deals with sustainable affordable housing in developing countries. In particular, the research aims to deliver an integrated development plan for a new business ecosystem design model oriented to new sustainable settlements in Mogadishu (Somalia), involving local entrepreneurship, social facilities and renewable energies. Indeed, the topic of sustainable affordable housing in developing countries is gaining increasing importance for Somali and international stakeholders. Nowadays, the major gap in the provision of adequate and affordable housing is to build a social community and to go beyond just providing basic shelters, to create sustainable durable settlements. The fragile and uncertain nature of the social, political and economic context, characterized by the lack of common shared legislative references and business strategies within the housing sector, makes Mogadishu a complex and challenging reality to be explored and improved.

Keywords Sustainable housing settlements · Circular economy · Affordable housing · Mogadishu/Somalia/African countries

1 Background of Mogadishu Application Context

The research focuses on Somalia, a country which entails significant challenges and, despite decades of international involvement (UN-HABITAT 2008), is still placed among the five least developed nations according to the ranks on the 2012 Human Development Index (UNDP 2012). In particular, the extent of the project is narrowed to Mogadishu, the capital of Somalia where all the challenging, interrelating problems and opportunities are noticeable.

After the Civil War of 1991, Mogadishu was characterized by a period of instability where abuses, destruction and security issues led the majority of the population to

---

O. E. Bellini (✉) · A. Campioli · C. Del Pero · C. Talamo · N. Atta · A. Dalla Valle
Architecture, Built Environment and Construction Engineering—ABC Department, Politecnico di Milano, Milan, Italy
e-mail: oscar.bellini@polimi.it

N. Aste et al. (eds.), *Innovative Models for Sustainable Development in Emerging African Countries*, Research for Development, https://doi.org/10.1007/978-3-030-33323-2_1

escape from the city—since the war left massive devastation of houses and infrastructures—looking for better living conditions (Grünewald 2012). In the stream of time, the continual presence of no stable government in Somalia has caused a real people suffering and overall poverty, along with the citizens' displacement (Ahmed 1999). In this way, political factors strongly affected the social and economic development of the city with relevant implications for the whole Somali country.

Today, the situation has improved but it is still distinguished by poverty, social and political instability, shortage of basic services and institutions, as well as pervasive insecurity (Aisen and Veiga 2011). Hence, citizens are more and more expressing their need for ransom. As an evidence, currently Mogadishu is characterized by economic recovery, reconstruction and by a strong sense of optimism to the point that many Somalis residing outside the country are returning to their motherland (Aisen and Veiga 2011). However, it is important not to overlook that the concurrence of these factors could cause a sudden and uncontrolled growth with a consequent risk of speculation as well as purely economic strategies that do not take into account social and environmental sustainability aspects. Moreover, since there are no observatories and agencies that collect data on the financial situation in Mogadishu and in Somalia, this possible accelerated growth and the related negative effects may not be easy to recognize, track and analyse (Webersik 2006). In addition, although it has been at the centre of humanitarian operations in the twentieth century, it is possible to observe how recently Mogadishu is basically absent from the debates on humanitarian aids in Somalia (Grünewald 2012).

In this context, the BECOMe project focuses on the housing sector with the aim to mitigate the risk of the occurrence of this accelerated growth, proposing alternative models in line with the Sustainable Development Goals (UN 2014) promoted by the United Nations in the Agenda 2030 (e.g. Goal 1—end of poverty; Goal 3—good health and well-being; Goal 7—affordable and clean energy; Goal 11—sustainable cities in communities).

Concerning the housing sector, the political instability has limited the development of an affordable social housing and the protracted conflicts that have gradually destroyed the local architecture, engineering and construction (AEC) firms as well as the manufacturers of construction materials and components (IIED 2019). Indeed, the commercial policies strongly influence the building-related choices (Davies 1987) and induce the operators of the sector to opt for the cheapest choice even if it is not the most effective one. Furthermore, the sector has been influenced overtime by the intense growth of population, the high levels of poverty, the destruction of the building stock, the displacement of population and the insecurity of the area as result of Civil Wars (UNFPA/PESS 2014), which has increased the housing demand (Chirisa and Matamanda 2016). However, nowadays, the construction sector is on the rebound exclusively for what concerns the high-level housing but not for the affordable housing, whose demand still remains unsatisfied (IOM 2018). Indeed, the business dynamics concerning low- and medium-level housing are highly affected by the context specificities and the uncertainty conditions that discourage both local and foreign investments. Moreover, there are two major additional barriers for the improvement of housing sector. Firstly, Mogadishu is characterized by widespread destruction of infrastructures (hard connection) and of supply chains (soft connection), posing challenging issues for rebuilding programs. Secondly, the unemployment rate is in overall terms higher than 60% (CAHF 2018), and the per capita income is around US$435 (Altai Consulting 2016), leading to a low purchasing power.

In this complex and challenging scenario, it is ever more recognized the necessity to promote and support a sound and conscious growth towards a new sustainable affordable housing development.

2 Interpretive Hypotheses for Sustainable Development of Mogadishu

The selection of Mogadishu as a specific application context is due to the need to provide informative and interpretative supporting tools useful to mitigate the risk of a sudden accelerated economic growth in the construction sector—operating according to undefined commercial policies—that may prevail over environmental and social issues. Moreover, the construction sector currently focuses only on high-level housing without involving the low and medium levels. To face the challenges that characterize the reference context of Mogadishu towards sustainable development, the research pursues the three pillars of sustainability. Hence, it is based on the following interpretive hypotheses:

- the development of a model of housing that integrates spaces for local crafting/manufacturing activities, ICT development environments (learning by sharing) and social services as key factor to facilitate, support and stimulate the local micro-entrepreneurship of Mogadishu;
- the integration of on-site renewable energy production (at no cost for housing units and at an affordable cost for local entrepreneurs and social and business services) as an enabler of optimization and innovation of energy production and management practices, as well as the promoter of new sustainable and high-performance technological solutions;
- the revamping of the local building sector with the manufacture of construction materials/components and the creation of new appropriate supply chains and training strategies as a means to make the most of local resources and boost the local construction sector.

It is worth mentioning that these interpretive hypotheses must be validated and adjusted (if needed) according to the results of a preliminary investigation of the specific context of Mogadishu, taking into account political, legislative, economic, financial and social aspects.

3 Goals and Objectives Towards Sustainable Affordable Housing in Mogadishu

In this context, the research project aims to propose a new business ecosystem for sustainable settlements, developed through an integrated model that embraces affordable housing, local entrepreneurship and social facilities, also exploring the exploitation of local loops in a circular economy approach.

Therefore, the research goal is to outline

- a set of possible scenarios able to stimulate new investments within the building sector, ensuring a balance between all three pillars of sustainability.
- a methodology for evaluating, for each scenario, feasibility conditions (economic, legislative, social), assessing direct and indirect benefits and risks.

In particular, from the social point of view, the proposed affordable housing model is designed to target a pre-defined housing price that can be affordable for the 70% (at least) of the population, through the optimization of the entire construction process and the use of the local renewable energy. From the economic and financial point of view, it suggests an investment plan, leveraging financial aids for local entrepreneurs and renewable production as well as showing the potential for sustainable intervention by external investors. From the environmental point of view, it strives to offer a high energy performance of buildings and consequent adequate comfort levels through climate-responsive design and the exploitation of most appropriate construction materials and techniques.

In addition to these three main sustainability goals, BECOMe has a broader purpose; the research involves the use of bottom-up actions to increase in the new generations the civil maturity and responsibility in order to overcome the ethical, economic and environmental challenges of the future. At the same time, BECOMe aims to raise awareness of the local government on the potential for providing affordable housing solutions to a larger part of the population, creating—through the involvement of local entrepreneurs—a lively and almost self-sustaining environment and local community.

4 Methodology to Deal with the Fragile, Dynamic and Uncertain Nature of Mogadishu

The research follows a multidisciplinary approach involving the deep interaction among four key disciplines, considered as fundamental for developing the proposed business ecosystem of sustainable settlements adopting a holistic perspective:

- architectural design (typological/functional aspects and the relation with the context);
- building technology (construction technologies, process organization and supply chain management);

- energy and building physics (energy design and the integration of renewable energies);
- management engineering (business ecosystem models and their economic sustainability assessment).

The research addresses the business ecosystem in its entirety, framing the complexity of the project from several points of views, but focusing, in this paper, on technological and productive aspects.

The development of the methodology derives from the awareness of the specific social and political conditions in Somalia characterized by a very fragile, uncertain and dynamic nature. The specificity of Mogadishu context, with respect to the research, involves issues of different nature related to the difficult collection and retrieval of data, the lack of certain and up-to-date references and sources, as well as the ambiguity in reading the political, legislative, financial, social and productive contexts. Given the difficulty of acting in such a fragile context, the proposed methodology may change over time according to the results of each phase of investigation and experimentation.

On the basis of these premises, the project will be developed following a methodological approach according to different interconnected phases that may be adjusted according to the partial results of each phase. In particular,

- the first investigation and analytical phase is based on the searching, collection, assessment validation and processing of data, related to the various characteristics of the context.
- the second phase identifies the needs and requirements (regarding the integration of housing, working activities and services) and relates them to the available resources (skills, funding, production capabilities), which can be activated both at the local level and at the international scale. The result is a settlement model based on strategies and rules regarding appropriate/appropriable urbanization, typological schemes, integration of solar systems, production approaches, facilities, technologies and materials.
- the third phase aims at identifying leverages and barriers for the application of the settlements model in relation to various internal and external opportunities and risks connected within the context. In this phase, the research investigates various aspects (stakeholders' interests, funding forms, financial instruments, etc.) putting them in relation to the data collected in the first phase. The expected results are the development of business ecosystem hypotheses, outlined in relation to different context scenarios (settlement dimension, type of investors, forms of business relationships, financial instruments, etc.) and the elaboration of business relationships and plans.
- the fourth phase adopts the methods and tools of the risk assessment and of the PEST/SWOT analysis in order to evaluate both the feasibility and sustainability of the proposed scenarios and develop notes for guidance for the stakeholders of AEC sector and of the related industries, identified in the first and third phases of the research.

– the fifth phase deals with a series of actions aiming at disseminating and communicating partial and final results. In particular, the results of the first and third phases are useful to activate and/or support capacity building and the creation of enabling tools and environments. Besides, the development of training modules directed to different kinds of users is necessary in order to share a basic knowledge between the identified stakeholders.

From the first phase of investigation of the application context of Mogadishu, it emerges the technical difficulty in finding updated and validated information, able to describe the applicative context. The absence of information tools and consolidated interpretative instruments, the presence of uncertain legislative framework and institutional systems imply epistemological issues that impose the formulation of an ad hoc research method, open to modifications during the research development.

Nevertheless, the proposed methodology is expected to be highly scalable to other similar application contexts. Indeed, given its uncertain, fragile and dynamic nature (lack of infrastructures/technical skills, political instability, etc.), Mogadishu can be considered as a complex application field that allows to envision the replication of the proposal in other realities (African countries) as a downgrade in complexity.

5 A New Business Ecosystem to Boost Mogadishu Housing Sector

The core goal of the research consists in designing and proposing a business ecosystem suited to the social/economic context and issues, able to activate a new housing market and to attract and engage different stakeholders. For this purpose, the research defines the conditions of pre-feasibility of the proposed business ecosystem, and it suggests the methodological framework and tools for drawing and assessing possible scenarios of development.

Starting from this main objective, the project results are as follows:

1. a model of modular settlements, to integrate low-cost houses, business units for artisans/small local enterprises and social services. The settlements are characterized by

   – pre-defined housing prices affordable for at least 70% of the population;
   – typological schemes appropriate to the local culture;
   – integration of photovoltaic systems to ensure reliable and affordable energy access.

2. a set of scenarios of possible actions, related to the development of the business ecosystems on large scale, aiming to create local enterprises and to stimulate foreign investors for the revamping of the national AEC sector and of the related industries. The scenarios are methodologically defined by linking in a matrix form the different topics, such as the production approaches (from artisanal and

traditional production to industrialized processes), the types of construction materials, their sources (local resources, resources from outside, recycled local war debris), the types of building components, the local and foreign stakeholders and the forms of business relationships;
3. a methodology for evaluating, for each scenario, economic feasibility conditions, through assessing direct and indirect benefits and risks. Financial instruments (e.g. mortgages and loans) and available financial aids programs supporting economic feasibility have been investigated.

Up to date, concerning the technological and productive aspects, the research is still at an early stage of development facing—starting from the very beginning—difficulties in the collection of updated, reliable, consistent data, information and documents. This lack of consolidated sources and references implies a significant effort in the identification and definition of the specificities of the application context. The information shortage and inconsistency lead to develop a strategy to engage key stakeholders that may affect or be affected by the proposed affordable housing model. In particular, from the results of the stakeholder analysis, the following four macro-categories are identified:

- *Private stakeholders of the AEC sector*: (a) Local and international construction and manufacturing firms, SMEs, medium/small social cooperatives, developers; (b) Local and international architects, engineers, builders, etc.
- *Institutional stakeholders and NGOs*: policymakers, national and local authorities (Government, Municipality of Mogadishu, etc.), NGOs.
- *Investors and donors*: private/public investment funds.
- *Citizens:* displaced people from Mogadishu, middle and lower class, workers looking for new opportunities.

From the dialogue (interviews, brainstorming sessions, e-mail correspondence, etc.) with these key stakeholders, it was possible to extract useful information concerning the recurring technological solutions. In particular, the most used solutions (i) for the structural systems are the reinforced concrete pillars; (ii) for the wall system are the hollow cement blocks; (iii) for the slab system are the full reinforced concrete slabs; (iv) for the roof system are the wooden trusses on corrugated galvanized sheets. This first exploration on the recurring technological solutions allows to concentrate the investigation on the production and supply chains of these products and materials. In this regard, actually in Mogadishu, it is possible to observe few attempts of supply chains development that involves both local and international stakeholders (e.g. Turkey, China, United Arab Emirates, etc.) for the provision of the above-mentioned technological solutions.

In addition, the first research results show a very challenging picture of Mogadishu for what concerns the construction sector, firstly, in terms of lack of sensitiveness about high-performance technologies (the common practice is to opt for the easiest and cheapest solutions without considering performance and sustainability issues). Secondly, there is a proven complexity in mapping the production and supply chains both at local and international levels, due to the difficulty in finding updated, reliable and shared data on the application context.

These initial results must be taken into account as a knowledge base for further phases of the research project, conceived as an iterative process continuously fed by the new retrieved information.

6 Social Impacts for Mogadishu Local Community

BECOMe is purposively designed to have an impact on three different but interrelated players as follows:

(i) the population of Mogadishu;
(ii) the local Architecture Engineering Construction (AEC) sector;
(iii) the local entrepreneurs (cooperatives, social enterprises, etc.).

More in details, by proposing solutions for meeting the demand of low- and medium-level housing, the project addresses the needs of the local vulnerable segment of the population. In particular, on the one hand, the population that currently lives in shelters could benefit from the opportunity to move into houses that are no longer improper and temporary but adequate and durable, increasing their conditions of hygiene, safety and well-being and, in general terms, ensuring a higher quality of life. On the other hand, the population that presently resides in low-level housing could have the opportunity to upgrade their conditions moving into medium-level housing, with the chance to integrate micro-entrepreneurship spaces as a support for their business development.

In addition, the actors of the local AEC sector may strongly benefit from the definition of a model of modular settlements, where different alternatives and chains of supply have been already analysed and grouped. This "plug and play" model will help them in the evaluation of practical business opportunities and in establishing tighter relationships along the supply chains. The potential for creating a circular economy approach exploiting the local presence of specific materials has been also explored. Specific training sessions and materials have been developed by BECOMe team in order to ensure a proper understanding and practical utilization of the model.

Moreover, BECOMe project has the aim to provide an impact throughout the entire life of the settlements. Indeed, the idea of creating a local community of entrepreneurs, offering services to the inhabitants of the settlements and to the local community is one of the key point of the proposal and has brought in the project the need for adding some specific facilities, such as—among others—the office/SMEs dedicated spaces in the buildings and the renewable energy production plants.

Local entrepreneurs can benefit from the above-mentioned facilities and from a dedicated training and documentation aimed at helping them understanding the potential for the use of such facilities in their businesses. The emergence of local services for the support of the life cycle of the settlements, such as electric maintenance and refurbishment construction materials supply, has the potential to create a virtuous circle helping the settlements becoming an almost self-sustaining ecosystem within the local community.

7 Conclusions

BECOMe project proposes a new business ecosystem model for sustainable settlements, integrating affordable housing, local entrepreneurship and social facilities, also exploring the exploitation of local loops in a circular economy approach. The project takes into account the fragile, uncertain and dynamic nature of Mogadishu application context, facing—among others—issues concerning the difficulty in data acquisition, the lack of reliable and shared references. In this regard, the project proposes new informative and interpretative supporting tools useful for both foreign and local investors, allowing them to gain a comprehensive and holistic view of the real conditions, opportunities and barriers of Mogadishu context. Indeed, the project can help foreign and local investors in the assessment of business opportunities, providing valuable context-aware information and indications useful for the development of pre-feasibility and sustainability plans for investments.

Lastly, BECOMe supports the research community by studying in details the practical applicability of a model of modular and replicable settlements in a high-risk environment like the one of Mogadishu. The replicability of the developed models and tools of analysis in other high-risk environments, e.g. in other African countries, could open interesting avenues of research. Furthermore, the same multidisciplinary approach used in BECOMe applies in this respect to different research streams: from circular economy approaches applied to buildings in war contexts, to sustainable affordable housing design and construction materials production.

References

Ahmed II (1999) The heritage of war and state collapse in Somalia and Somaliland: local-level effects, external interventions and reconstruction. Third World Q 20(1):113–127

Aisen A, Veiga F (2011) How does political instability affect economic growth. IMF working papers WP/11/12

Altai Consulting (2016) Youth, employment and migration in Mogadishu, Kismayo and Baidoa

CAHF—Centre for Affordable Housing Finance in Africa (2018) Housing finance in Africa. A review of Africa's housing finance markets

Chirisa I, Matamanda AR (2016) Addressing urban poverty in Africa in the post-2015 period. Perspectives for adequate and sustainable housing. J Settlements Spat Plann 7(1):79

Davies R (1987) The village, the market and the street: a study of disadvantaged areas and groups in Mogadishu, Somalia

Grünewald F (2012) Aid in a city at war: the case of Mogadishu, Somalia. Disasters 36:S105–S125

IIED—International Institute for Environment and Development (2019) Accessing land and shelter in Mogadishu: a city governed by an uneven mix of formal and informal practices

IOM—International Organization for Migration (2018) Shelter projects. East & horn of Africa: 14 case studies

UN (2014) The millennium development goals report

UNDP—United Nations Development Programme (2012) Somalia human development report 2012: empowering youth for peace and development
UNFPA/PESS (2014) Population estimation survey 2014 for the 18 pre-war regions of Somalia
UN-HABITAT (2008) State of the world's cities 2010/2011-cities for all: bridging the urban divide
Webersik C (2006) Mogadishu: an economy without a state. Third World Q 27(8):1463–1480

New Foundation Cities

Massimo Ferrari, Claudia Tinazzi and Annalucia D'Erchia

Abstract As part of the Italian tradition related to urban projects, some experiments throughout the twentieth century have shown, in the comparison between the different possibilities, the specific ability of architecture to lend concrete form to the living environment of a civilisation in a specific age. The design of the city, of its way of expanding, has outlined, in the succession of examples built or even only conceived on paper, the possible prerequisites for the definition of some principles aimed at the determination of sections of the city, or in the most virtuous examples of new foundation cities. Every latitude, just as every epoch, enjoins in this sense the need to re-examine these principles, which, if on the one hand express universal and timeless values, on the other hand, search for increasingly greater relevance to specific cultures as well as to needs and demands associated with one's own time. Africa's living future rests on a recent past already quite rich in experiments, on a founding custom that in the previous century has built new urban centres, capital cities, transfers of centrality to regular federations of states. Living in Africa, besides contemporaneity, represents from this point of view the most extreme modernity; living consistently with the culture, history and traditions of a country that has forever portrayed in Western imagination no more than the mystery and exotic dream of a continent that is still unknown if not actually stigmatised in its most conventional characters.

Keywords Urban projects · New city · Le Corbusier · New leaving · Tropical belt

1 Imagining the Future in Africa

> [...] Inside Tibesti, an indigenous guide asked me whether by any chance I wanted to see the walls of the city of Anagoor, as he would have accompanied me there. I looked at the map but the city of Anagoor was not there. Not even on the guides for tourists, normally so rich

M. Ferrari (✉) · A. D'Erchia
Architecture, Built Environment and Construction Engineering—ABC Department, Politecnico di Milano, Milan, Italy
e-mail: massimo.ferrari@polimi.it

C. Tinazzi
Milan, Italy

N. Aste et al. (eds.), *Innovative Models for Sustainable Development in Emerging African Countries*, Research for Development, https://doi.org/10.1007/978-3-030-33323-2_2

> of details, was there any hint to it. I asked, 'What kind of a city is this that is not marked on the geographical maps?' and he replied, 'It is a large city, extremely rich and powerful, but the geographical maps do not indicate it as our Government ignores it, or pretends to ignore it. It fends off for itself and does not obey. It lives independently and not even the kings' ministers may enter it. It does not trade at all with other countries, whether near or afar. It is closed. It has lived for centuries inside the circle of its solid walls. Does the fact that none has ever exited it not mean perhaps that they live happily there? [...] (Buzzati 1958)

As an exemplary retribution, an unexpected vision built by a fervid imagination, the happy discovery of the city that is not, in the thirty-ninth short story out of the sixty Dino Buzzati wrote immediately after the Second World War of the twentieth century, urges us to research—by contrast—the concrete possibility of inhabiting an ideal city imprinted precisely in the African territory that has provided the background of the story as well as our utopia while reading the text. Written Africa, in the literature of any country in the world and in Italy depicted more recently by Ungaretti, by Marinetti, and then by Manganelli, Celati and before them Moravia, with all the suggestions conveyed to the artist and architect friends and shortly before him by Bianciardi—as true as the sour Italian life—betrays in the anxiety for truth every exotic dream or, by contrast, shrinks down sometimes to folkloristic cultural stereotypes and exits the simple and measurable reality to appear as a mirage brimful of qualifying adjectives like the beginning of this text.

The walls of Anagoor are an invisible possibility of living out Africa, a precise limit between being inside or outside a place. They represent a way perhaps too western to imagine a space to inhabit, fruit of the subconscious of our civilisation that retains these high walls in the roots of its own history; they are—still—the naive demonstration of superimposing one's peculiar reality on other cultures despite the recognised ability to the surprise of a special writer like Buzzati, who throughout his opus has entertained a happy relationship with the dimension of lived space, translated into a constant process of rarefaction and abstraction within his stories.

Founding a city, imagining its construction within a single time span, even in the circumscribed hypothetical scenario, as in the precious character of Anagoor, might possibly mean to image first of all a way of living free from an excessive weight of the memory of past forms, a way contemporary to the current age, suited to the host place, suited to the general conditions of entrenched nature as well as to the special qualities that are read on the smaller scale. It means—still—to face historical and cultural anthropological diversities so that their inequality can shape the forms and the distribution of new city sections, to imagine shared principles, rather than predefined forms, criteria capable of ordering and highlighting the nature typical of the territories and the needs of those inhabiting them, to translate the habits into precise and recognised spaces and ambitions into new places.

The future of Africa, never so close as in our age, must, because of this, face the most genuine and radical features of a land that for too long has been viewed solely as a ground for conquest similarly to many other parallel countries in terms of latitude, social history and quality of primary resources; at the same time, however, the research must necessarily avoid ascribing a protected role, frozen in conventions or portrayed in foreigners' images, fruit of a popular tradition directly proportional

Fig. 1 Fausto Melotti, L'Africa, 1966

to the physical distance from the black continent and reconsider Africa, instead, according to its peculiar qualities, needs and possibilities, just as any other inhabited place on the planet. Imagining, perhaps already in the intentions, a more consistent evolution in the way these places are inhabited speaks to us of the future (Fig. 1).

Another writing, this time by Giorgio Manganelli, almost impossible to find, clarifies this antipathetic position, far from myopic, and helps us freely read the possible comparisons with a territory forever in motion:

> [...] «Animals populate the African space as an emblem they are required to make intelligible. Not colosseums but lions, not towers but soaring giraffes, not acropolises but craters crowded with wild beasts» [...] (Manganelli 2006)

Africa in motion, a vast chessboard consisting of constantly migrating *live monuments* in lieu of habitual and familiar stone constructions, clarifies the idea of a territory it is difficult to come to terms with, an environment we need to know and interpret starting from the violence of the extreme conditions that have designed it and still continue to design it today (Figs. 2 and 3).

As written earlier, the reality of Africa has shown a founding custom transcribed into new urban centres, capital cities and transfers of centrality to regular federations of states. The colonial capital cities and later the federal independence represent the social and economic drive for these new designs that trace in the first half of the twentieth century, already since 1900, vaguely traditional urban centres, Western-style

Fig. 2 Romuald Hazoumé, Wax Bandana, 2009

Fig. 3 Yinka Shonibare, How to Blow Up Two Heads at Once (Ladies), 2006

cities in which the orthogonal matrix structure, with minimum diagonal variations in its distributional evidence, accompanies without contaminating the pre-existing indigenous forms: thus are born, for instance, N'Djamena in Tchad, a military city founded by the commander Émil Gentil, Abidjan in Ivory Coast, de facto capital even after the political power was moved to Yamoussoukrone in 1983, and later Kinshasa in the Zaire of the 1930s as well as Niamey in Niger in 1937. The end of the century even prior to the 1900s had laid the colonial bases for these foundations that are in actual fact re-foundations or continuations of long-term plans of occupation of the territory or further new expansions, precisely like Kinshasa vis-à-vis Brazzaville founded on the other side of the Congo river in 1881. Cities of the calibre of Bamako in Mali, Porto-Novo in Benin and Lomé in Togo, to mention but a few of them, mark the northern European challenge of the second half of the nineteenth century, an age of explorations and simultaneously of the slow decline of Western expansionist dreams. Their regulatory blueprints collide with the different dispositions of road networks arranged without any overall designs, centralities absent or disconnected from the fabric confine the urban relevance of these centres to no more than their distribution on the territorial scale, to the juxtaposition to connecting historical paths or to favourable inclinations to settlement along the coast.

Still a memory, one just elapsed, should be acknowledged to the North African experience; from the colonisation of French Algeria, occupied by approximately 600 small centres between 1836 and 1914, leaving aside the utopian and more famous visions for Algiers of Le Corbusier, to the realities built in Ethiopia, Eritrea and Libya through the Italian experience in the age of fascism, which without any second thoughts superimposed, on the African territory, small European matrixes. Metaphysical cities narrated by Gherardo Bosio's plans for Gondar and Dessiè and those for Asmara of Odoardo Cavagnari, the drawings of Ignazio Guidi and Cesare Valle define the character and geometries of Addis Abeba, of which it might be important to recall the earlier, freer and more modern general development plan drawn by Le Corbusier in 1936 and offered directly to the Duce; a unified design rediscovers the characteristic themes of Le Corbusier's philosophy, nature, residence, public places set up in an open structure, arranged through precise geometries, organised by the infrastructures that never found a concrete possibility of implementation. The very recent history delineates yet a different phenomenon in the design of these new centres, transfers, movements of capital that pursue a political logic associated with the governmental independence of some nations; displacements that from the second post-war period to the middle of the eighties built Dodoma in Tanzania, Abuja in Nigeria and Lilongwe in Malawi, anthropising, within the individual nations, strategic positions often limited to the perspective of political opportunities, replacing the previous orthogonal thread with a more organic geometry and yet ending up with the same inconsistencies.

The relationships between these hubs and the territorial connection networks, vital skeleton of any country, raise, perhaps without any mediation, a crucial issue for the African continent. The territorial structures that, without disregarding their obvious functional usefulness and the clear construction necessity, have mostly influenced over time the ancient landscape in its natural dimension are undoubtedly traceable, since the original epochs, to the infrastructural system and to all the possible variations and expressions of such a vast topic susceptible of being defined in its pivotal points: roads, bridges, aqueducts, without any hierarchical or chronological order, have first disclosed, in their rational habit, the features of necessity vis-à-vis a project of collective work capable of transcending the single idea of living (Fig. 4).

However, the traces, as we continue this diachronic journey that sums up different attempts at foreign anthropisation, have been imprinted on the African soil since epochs long before the recent past, starting from those crossed lines that the Romans

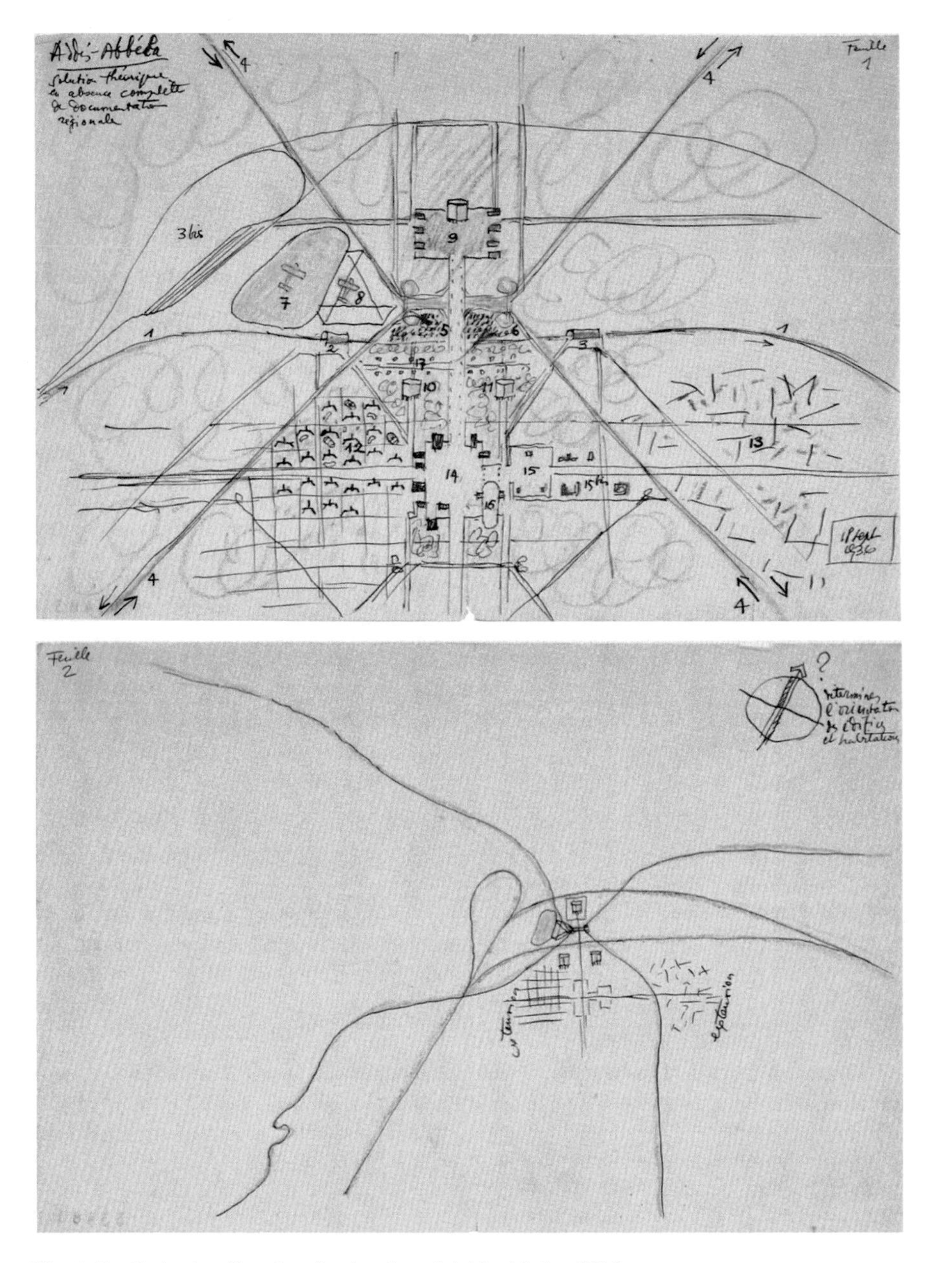

Fig. 4 Le Corbusier, Sketches for the plan of Addis Abeba, 1936

indelibly engraved on the sandy ground north of the continent, the structural system of Decuman roads, the residential typologies absorbed and revisited by subsequent periods and the variations on the collective places of the Forums and the Thermal Baths (Ferlenga 1990) which, precisely on account of their functional adequacy and the features of generality tailored to the environment, have been able to turn into a matrix of subsequent typological and urban developments, so much so that they could become successfully integrated and shape up in later eras the true constants and the recognised character of the inhabited environment preserved until our times.

After all, leaving aside the recent attributions to the system of journeys the Carthaginian admiral Annone seemingly undertook up to the African equator around the sixth century BC, the ancient indigenous reality marks, unlike these steady elements, all the transient characters of the systems of human settlement, from nomadism to transformation, from precariousness to the variation of orientation: exemplary in this sense is the physical rotation of each individual building inside the primitive villages to face the residence of the new chief after the death of the previous one; the sole constants might be set for a long time in the use of materials and the building techniques.

In this suspended space, we can place the difficult balance of contemporary research around living in Africa, which, moving from the highest imagination, must nevertheless entrench itself in a new awareness of the reality that considers without differences the territories long incomprehensible, borrowing a leaf from the words of Gianni Celati at the end of the last century:

> [...] Let us mull over the fact that by now anthropologists have hardly anything to do with primitive populations, reduced to degenerate tramps or exotic background actors. Some rare team chases after the last groups in the forests of Amazonia, but if they find them still naked with bow and arrows, they infect them at once with cold or influenza, lethal diseases for them [...]. (Celati 1998)

2 A Collective Research Laboratory

> [...] See, O future, I have mounted on your horse; what new banners are you raising towards me from the towers of cities not yet founded? What rivers of devastation from the castles and gardens I used to love? What unforeseen golden ages are you preparing, poorly mastered, you harbinger of treasures paid dearly, you kingdom of mine to be conquered, you ... Future [...]. (Calvino 1959)

The idea of future always gathers in the common imagination all the conventions that, without authentic reasons, crowd contemporary history and without any concrete logic produce relationships out of size with the historical ability to read changes and transformations. It is, however, typical of architecture, in its disciplinary custom, to launch and substantiate designs starting from parallel researches generated by social, humanistic and scientific disciplines without in any event ever transcending the horizon that from its viewpoint narrates the future. Living in Africa, beyond contemporaneity, means to search without any hypocrisy for the reasons that might represent the peculiar qualities, the dimensions, the timeframes and the needs of a settlement that have invariably evinced all the hardships of living in an extreme region in terms of a latitude and longitude, not just in a geographical sense. Settling in the territory ultimately means to forge an agreement with the environment that hosts us, capable of listening to the reasons of a nature often removed from our imaginations; it means thinking of the future as opportunity to know and accept differences and thereby live out the world coherently. Throughout the African continent, over the last centuries, lifestyles and traditional cultures have undergone profound changes in the encounter–clash with the modern age. The economic, political and social conditions have been altered first through the establishment of the colonial states and then through the creation of the independent autonomies, within a society internationalised in an increasingly worse manner.

We thought that one of the most concrete ways of imagining the future of this continent was to conceive of it as inhabited, ideally built, using the typical tools of our discipline to provide answers to the urgent contemporary needs, combat extravagant possibilities and propose new idealities, leaving the question open: showing a way of living in Africa. The work proposed is a collective research, not a celebratory exhibition of individual design skills; it is a choral work of engagement around a topic as urgent as ideal. A laboratory constantly in motion, experimental and virtual, which shows the need to supplement research, the precious critical work of reorganising both the recent and the distant past, with new design proposals, reflections that are not conclusive yet profoundly achievable. We thought of inviting some architects and design groups so that they could narrate to us their idea of settlement, free from any preconceived constraint, unburdened by any pre-existing grid or design, aware of the past but leaning towards the future (Fig. 5).

We thought that to know and inhabit the tropical belt, live out and interpret a unified geographical area, in many respects common to South America, Africa and Asia, might be a necessary premise to get consistent responses for the interpretation of a territory, of a landscape, of an environment as hostile as captivating. A choice detached from any economic or neo-colonialist logic, a selection respectful of concordant geographical units that in a certain sense unify a community of architects influenced by a habitual style of living; a geographical identity, the tropical one, still

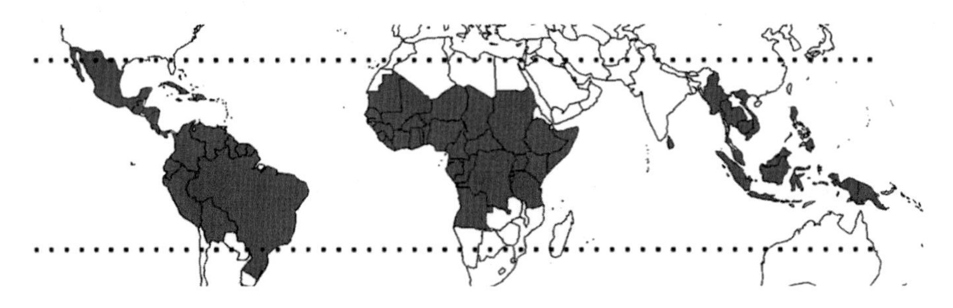

Fig. 5 Tropical strip in which the architects were chosen for future city projects

to be investigated in depth, to be ascertained through a reading capable of enlarging the customary geography and smashing several prejudices, to find new and different forms of mapping more influenced by the climatic bands, by the styles of living, by the common cultures rather than by the political limits of regions, states and nations whose certainty exists only on paper. We thought that one of the qualities for selecting the proposed architects could lie in the fact that they had not been involved at international level with territorially large-sized designs and had built consistent architectures with a clear interpretative logic, acknowledging to the invitees a pronounced critical–compositional ability to put to the test on this extended occasion. We thought that an additional selection criterion might involve the generational sphere and that, accordingly, the authors invited might be identified within that band of intermediate age, younger, who in our view currently has the chance to lend an unexpected contribution to the topic of the future architecture. In many instances, the architects belonging to this geographical area have been trained in Europe or in North America with the desire at present to rediscover their roots transposed into a profoundly modern spirit. Architects chosen without any intended election, far from each other in terms of distance and yet, we believe, mutually close in terms of cultural sensitivity and geography, capable of representing today the possibility of a sincere discussion in which multiple ideas of future cities can be tested and verified (Fig. 6).

We asked each architect to briefly show his interpretation of a possible way of inhabiting and living the future in Africa, providing his own design of a future city free from an enforced geographical positioning or from predefined settlement quantities, a reflection and an urban structure loaded with ideality, references and suggestions steeped in the specific African reality (Fig. 7).

Bom architecture, Gabinete Gabinete de Arquitectura + Solano Benítez/Laboratorio de Arquitectura + Javier Corvalán/Taller E, Sebastian Irarrazaval Arquitectos, NLE, Mariam Kamara, Anupama Kundoo Architects,

Fig. 6 Triennale di Milano, The exhibition “Africa Big Change Big Chance”

Fig. 7 Triennale di Milano, The exhibition "Africa Big Change Big Chance" with models and research notebooks "imagine the future of Africa" (Albrecht 2014)

Boubacar Seck are but the first 7 architectural firms that along with us have concretely imagined new ideas for living Africa's future.

References

Albrecht B (2014) Africa big change big chance. Ed. Compositori, Bologna
Buzzati D (1958) Sessanta racconti. Arnoldo Mondadori Editore, Milan
Calvino I (1959) Il cavaliere inesistente. Einaudi, Turin
Celati G (1998) Avventure in Africa. Feltrinelli, Milan
Ferlenga A (1990) Africa. Le città romane, Clup, Milan
Manganelli G (2006) Viaggio in Africa. Adelphi Editore, Milan

Development of Social Welfare Architectures in Marginal Areas of Sub-Saharan Africa. The Case Study of the Gamba Deve—Licoma Axis in Mozambique

Domenico Chizzoniti, Monica Moscatelli and Letizia Cattani

Abstract The research concerns a strategic project designed to implement a service supplying system to rural areas of Sub-Saharan Africa through testing of cultural and social welfare architectures that meet the principles of local building and representation principles and the versatility and contextual characteristic of the site. The study analyses a possible approach that may be generalised for use in other contexts featuring high social marginalisation, focusing on rural areas crossed by the street axis Gamba Deve–Licoma in Mozambique.

Keywords Sub-Saharan Africa · Typological and figurative aspects · Social welfare structures · Prototype

1 Introduction

This essay is part of an experimental research project carried out by the Architectural Design Laboratory—(ADL),[1] coordinated by Domenico Chizzoniti, started with an

[1] A facility dealing primarily with research has supported all the department's architectural design activities at the various application scale. A wide range of research activities in the descriptive experimentation for the critical modelling of the architectural space related to the composition's discipline have been undertaken. In this area, several studies are underway in collaboration with other Italian and foreign research partner institutions. Research is ongoing on prototyping of emerging facilities in poor and marginalised contexts in partnership with Not-For-Profit organisations and associations to prepare experimental building models with social welfare functions. Website: http://www.abc.polimi.it/it/laboratorio-dabc/adl-architectural-design-laboratory/.

D. Chizzoniti (✉)
Architecture, Built Environment and Construction Engineering—ABC Department, Politecnico di Milano, Milan, Italy
e-mail: domenico.chizzoniti@polimi.it

M. Moscatelli
Lecco, Italy

L. Cattani
Fidenza, Italy

N. Aste et al. (eds.), *Innovative Models for Sustainable Development in Emerging African Countries*, Research for Development, https://doi.org/10.1007/978-3-030-33323-2_3

expression of interest from doctors with Africa Cuamm organisation in 2012. Following contact with the Department of Civil, Environmental and Mechanical Engineering—DICAM at University of Trento about the Gamba Deve–Licoma axis, the proposal has been deepened through a social and cultural integrated service system in the architectural design in developing countries laboratory coordinated by Letizia Cattani with Monica Moscatelli.

The objective of the following research project is to design an architectural prototype able to respond to medical, social and humanitarian emergencies in countries struck by natural disasters, war and situations of poverty. Currently, in the migratory emergency phenomena involving European contexts, just as in Third World countries, the provision of essential services and support, in terms of health care and food and water supplies, is entrusted to government organisations and NGOs such as the World Health Organization (WHO), International Bank of Reconstruction and Development (IBRD) and the United Nations International Children's Emergency Fund (UNICEF), committed to responding to emergencies relating in particular to weaker groups at high risk of marginalisation, primarily women and children.

Under these circumstances, the structure in place today is based on building criteria and methods that suffer due to their precarious, transitory and temporary nature. This feature, particularly to temporary works (from tents to mobile structures), makes them unable to cope with the development of emergency conditions so that they can only work for a short and limited period of time. The development and stability of irregular migration and especially the continual changes to admission processes and policies, not only in European countries, call for a practical reflection on how to support the social and health conditions of some emergency situations with a high rate of marginalisation.

The specific aim of this project was, therefore, to establish guidelines for a new emergency-equipped system (health, social and welfare) capable of meeting specific typological, representational and local construction requirements. The configuration of a system with such characteristics was operationally oriented by on the conceptualisation of the prototype and designed to evolve over time, adding to modular elements to the minimum unit depending on the application site and other parameters, such as climate, the local materials and user needs (Chizzoniti et al. 2014).

The guidelines for definition of the prototype are based on certain principles such as functional independence, so that each part of the prototype may be perceived as autonomous and used for different functions; flexibility and adaptability to achieve a system that can be composed in many different ways, creating different spaces for different functions (Falasca 2000); ease of transport to ensure the reversibility of the prototype so that it can be easily reconfigured and transported in suitable containers by water or on land; the aggregative abilities to permit assembly with other modules, resulting in various combinations; finally, ease of construction for manufacture of a product in which each individual part must be very light and made from insulating materials, heat and wear-resistant and easily reproducible materials (Novi 1994).

2 The Case Study of the Gamba Deve–Licoma Axis

Among the many responses associated with the emergencies, this research focuses on social welfare assistance in a rural district of Sub-Saharan Africa, and in particular, along an axis of primary importance in the district of Caia in Mozambique, known as the Gamba Deve–Licoma axis.

The research conducted by the DICAM at University of Trento[2] was our starting point for configuring a prototype system through appropriate examination of the types of building, the distribution choices and the figurative characters and the construction choices characteristic of the local culture.

Based on the design guidelines provided by the PDUT *Plano de Uso da Terra* (District Land Use Plan) and the main territorial planning tool used in the Caia district, in the centre of Mozambique, drafted between 2011 and 2012, the research aims to address a specific aspect of the more general topic of medical and social emergencies, by analysing the main infrastructural axis in this setting, the *Estrada Distrital No. 1* (ED1), which connects the centre of Gamba Deve to the east, with Licoma to the west (Fig. 1).

In close proximity to this route, approximately 45 km in length, small groups of building units are distributed. Within these, there is an insufficient and irregular allocation of services such as markets, mills, small health centres, schools and wells, which are entirely inadequate to meet the requirements of the population. The aim of this research is to identify a coordinated system of small interventions using existing resources as the main force of a development that recognise, in architectural identity, first and foremost, the building factor.

The latter is traditionally based on a population that is distributed according to its needs and possibilities, in *Vilas*, provision centres for the most important services and administrative sites; in the *povoados*, small community centres organised mainly around primary services (the well or the school) and, finally, in *mudzi*, single-family buildings located a few hundred metres from one another.[3] In accordance with national policies, the network of primary services in the Caia district has mainly developed since the end of the Civil War and, as with all rural districts in the country, public services are mainly schools, healthcare units and wells for water supply.

[2]The research activity focuses on urban and land planning. Specifically Isacco Rama, former associate of the Consortium of Associations with Mozambique—CAM, through his degree thesis (2014) supervised by Professor P. Bertola and Professor C. Diamantini, developed a project proposal for implementation of the *Plano Distrital de Uso da Terra*, PDUT (District Land Use Plan) for Caia offering the district a tool for improving the service delivery system along the Gamba Deve–Licoma axis.

[3]Data relating to the population distribution is contained in the PDUT—Plano de Uso da Terra Caia (District Land Use Plan) 2012.

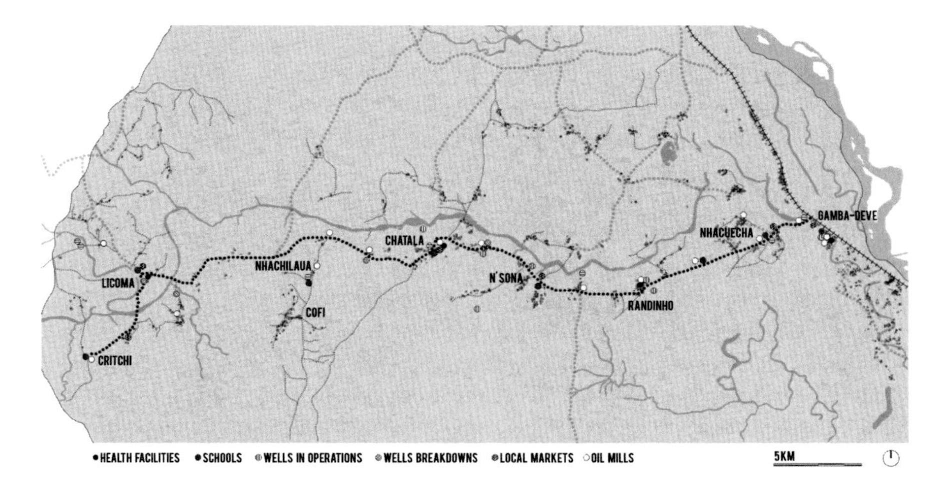

Fig. 1 Service supplying system along the Gamba Deve–Licoma axis

Alongside these, there are some activities run by private companies, such as the mills (electric or diesel-run) and the markets. The coverage of the services is not homogeneous in the district territory, and it is mostly concentrated in the more densely populated areas, in proximity to the *Vilas*, the major axis and the infrastructural networks between Caia and Sena; in rural areas, on the other hand, where housing dispersion in the area is highly dispersed, the network of services, whether they are public or private, is unable to satisfactorily cover the needs of the population. Moreover, the theme of the relationship between the city and the countryside, as well as the evolution of the building pattern associated with these, is fundamental to understand the socio-economic dynamics of the area studied. Indeed, while there is never a clear dualism between urban, structured and settled centres and the surrounding countryside, it is common to find simultaneous and mutually interfering development of the two realities which have a strong tendency to overlap, with the city often being a large rural settlement and the countryside taking on the typical infrastructural characteristics typical of a city.

For this reason, the revival of certain local morphological features, such as the organisation of rural areas in clusters, permits consideration of each centre as a place for trade support services, support for agricultural production, distribution and preservation of agricultural and livestock products, as well as a health centre where the minimum provision of assistance, which is guaranteed in every area, work alongside organised structures found only in larger centres.

In this traditional organisational system markets, schools, health centres and wells, they not only constitute essential services to the life of the rural population but also become genuine community centres, around which the primary settlement is organised.

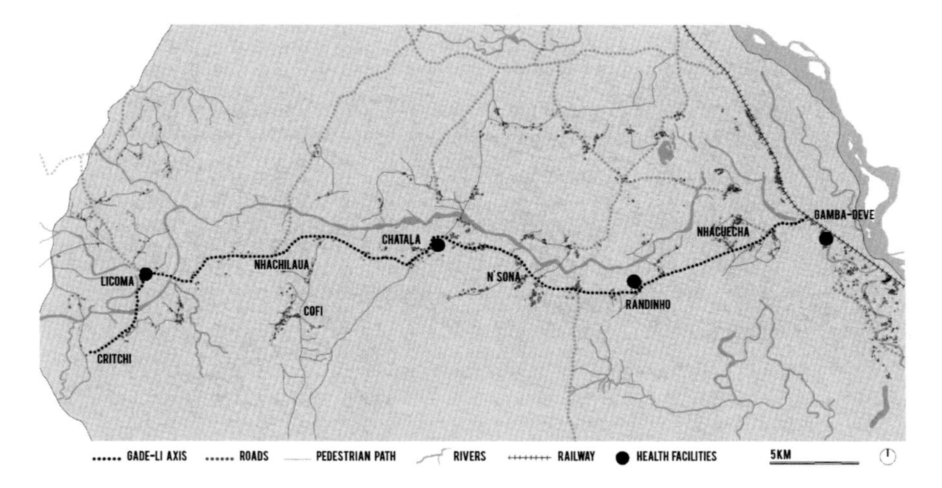

Fig. 2 Network of rural health centres along the Gamba Deve–Licoma axis

The local markets, for example, constitute for much of the population the only access to basic goods such as salt, soap, oil, clothing and local agricultural products. They are mainly located in the larger inhabited centres and in the area in question they are located in Gamba Deve, Nhacuecha, N'Sona, Chatala, Licoma and Critchi. As for the network of rural health centres (Fig. 2), this is organised on three levels: the *Posto de Saude*, a primary healthcare service designed to ensure an appropriate distribution of pharmaceutical and equipped to deal with the simplest of emergencies; the *Centro de Saude*, the intermediate circle of the rural health network, staffed by qualified healthcare personnel; finally, the *Hospital Rural*, a facility generally found in the district capital and divided into different departments, including surgery. Along the considered axis, the *Centro de Saude* (*health centres*) are located in Gamba Deve, Randinho, Chatala and Licoma, while the only structure in Randinho is a *Posto de Saude* (*smaller health centre*) (Rama 2013).

Similarly, the building pattern of the *mudzi*, either isolated or combined, meeting to the traditional housing needs of this specific context and identified as a building principle for new services too, together with the use of local resources and the involvement of the local population in the operational and constructive phase, outlines a strategy applicable to different areas, on a case to case basis, while at the same time maintaining a distinctive typological and representational form.

In particular, in this specific context, certain traditional building characteristics have been identified that reflect a specific usage of the land and space; indeed, apparently arranged at random, the *Mudzis* reflect, in reality, a very precise cultural and community hierarchy. Each such building is the result of entirely spontaneous activity, but has the immediate purpose of representing the size and importance of the household, basing its construction on the use of local material, which is easily available and affordable, such as wood, raw earth mixed with water, straw and bamboo.

Therefore, by refusing a fragmentary vision with sporadic opportunities for intervention and a conception of a large-scale (all-encompassing but sometimes abstract) transformation of the territory, this project adopted a working strategy for building units consistent with an appropriate scale where it was possible to find specific relationship of coherence with the scale of the natural landscape, linked to individual needs for urban or rural transformation with local morphological features and building characteristics. Thus, the various area of production and social reorganisation are revisited in this African context proposing coordination between ad hoc interventions and infrastructural reorganisation of urban areas and agricultural sectors, through a minimum provision of health and social welfare facilities.

3 Social Welfare Prototyping Experimentation

Recent experiments aimed tackling crises in countries affected by emergencies have relied on the development of a standardised quantitative model due to the urgent need to pre-empt critical situations minimising, however, the interest in the construction quality of the space, and therefore, robbing of it the compositional and figurative themes that define the architectural artefact through details. Here, we are seeking to understand how the production potential of modern industry may be evaluated in a critical standardisation process in which it is also possible to reconsider certain principles inspired by "quality in quantity" (Semerani 1978), in other words a more flexible production capacity to meet the need for provision of temporary constructions with greater structural complexity and, therefore, in the architectural terms, compatible, not only constructionally but also figuratively and environmentally and aesthetically with the circumstance in which we found ourselves operating.

The project addresses the compositional aspects that define the architectural space of the prototype derived from the investigation of the sociocultural aspects, the needs of the communities and analysis of traditional buildings (Neutra 1948). One of the fundamental aspects of this project is the union between form and materials, reflected in the design experimentation, in which form viewed as the act that precedes the design phase, is coherently expressed in its entirety.

Design of care, cultural and social structure requires a multidisciplinary approach in order to organise the building in a way that is over the time. The requirements in terms of the flexibility, modularity, aggregation and reversibility of the prototype are, therefore, a key element in responding to rapid changes and the needs of the population (Kleczkowski et al. 1985).

The prototype integrates health and social care characteristics with the economical, social, cultural and institutional conditions seeking exemplary social, functional and representational qualities within historical architectural experience.

To illustrate the proposed strategy, a multifunctional prototype was tested by a group of students who have worked with the ADL Group—Architectural Design

Laboratory.[4] The prototype, designed with modular elements easily adaptable to other very isolated contexts, consists of a simple low-cost structure easily implemented for services and health and social care provisions (Staib et al. 2010). The aim is, therefore, to create a multi-purpose structure as a focal meeting point for the local community in question.

The project defines its typological and representational choices through interpretation of the traditional architecture of the analysed context, namely the popular local dwelling (the *mudzi*), which consists of a group of small buildings (*palhotas*), arranged around a large circular open space (Nicchia 2011)—using the architect John Hejduk's studies of decomposition of space.

The flexibility of the space is obtained here through the decomposition of the a square (Fig. 3), the result of careful analysis of the complexity of Hedjuk's living spaces. Form and organisation of space, representation and reciprocity are elements of reflection for that author. In his three "Diamond House" designs, Hejduk is inspired by the paintings of Mondrian, using rotation of the square and on horizontal and vertical lines, which are transposed into his projects as dividing partitions. The partition becomes the modular element that makes up the healthcare and social spaces facility. These elements enable us to achieve considerable complexity, being arranged around a central space as the unifying core of the surrounding elements, a reference to the *palhotas* of popular local architecture.

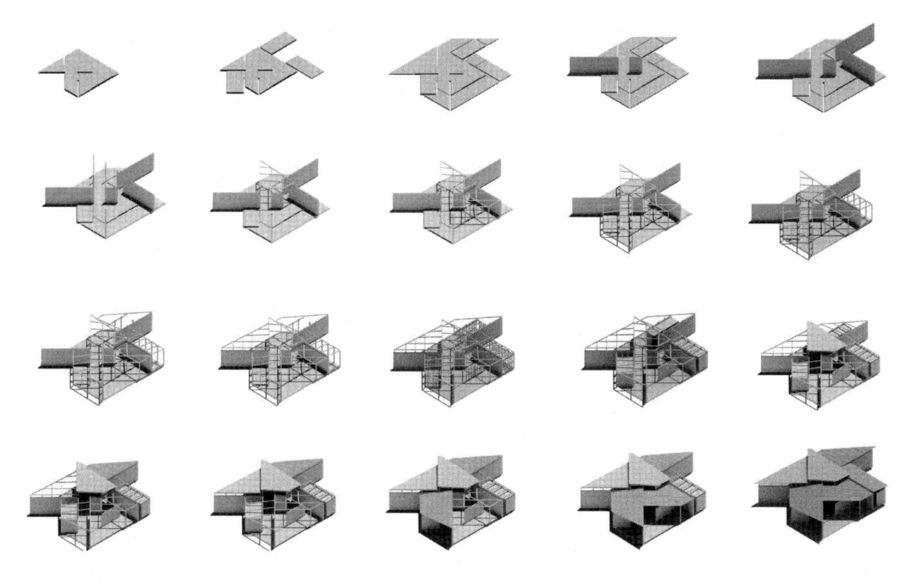

Fig. 3 Compositional process of the prototype through decomposition of a square

[4]The prototype was made in 2015 in collaboration with the students M. I. Guin, F. Martellono, F. Menici.

The roof, consisting of a pitched system supported by a modular beam that connects the various components of the prototype, reflects the complexity of the internal space. Thus, the system of hierarchies within the facility is highlighted, rendering the pivotal role of the central space, reminiscent of the *mudzi* clear and legible. The modularity of the prototype becomes the guideline for building the relationships between space and structure, defining the compositional process behind the project (Ceccarini 1989). The sustainability of the prototype is characterised by low construction costs, availability of construction materials and easy of construction. The load-bearing structure is made entirely of wood, a material that is easy to assemble thanks to its modularity and the partitions are covered with raw earth bricks that reinforce the structure (Fig. 4).

4 Conclusion

In conclusion of this journey, to return to some of the issues mentioned at the beginning, we would like to clarify the role of the versatility, reversibility and the temporary nature of the design, less in its authorial dimension than in a more conceptual one, and therefore, at least open to generalisation with regard to the various implementation scales; for example, the coherence between objectives and the means developed to implement them; between new figuration and contextual architecture; or even regarding construction and sustainable forms of incentives for the use of local resources and so on. This position is supported by design dimension that is not an exclusive finished product, but rather tends to open up its formalisation potential through a procedure oriented more towards prototyping than towards the author's model.

Guido Canella clarifies this position in his famous essay on building motivations and the experimental design of the Milanese theatre system. In his interpretation of architectural prototyping, Canella seeks to illustrate those features of experimentation with form, alluding to the prototype as a design synthesising distributive and geometric—and therefore also typological—conception tied to particular functional and social life system. He defines prototypes as "*educational projects*", indicating their instrumental aspect since they will occupy the final pages of some theoretical paper. However, the term "educational" perhaps also, unconsciously, signifies a self-critical reserve concerning the degree of simplicity, or if you will, approximation of architectural expression (Canella 1966).

The prototype is an architectural idea and must be equipped with a deep structure conception, as a part of the typological research. The prototype engages the limits of an early theoretical reflection, which precedes the actual act of design, contemplating

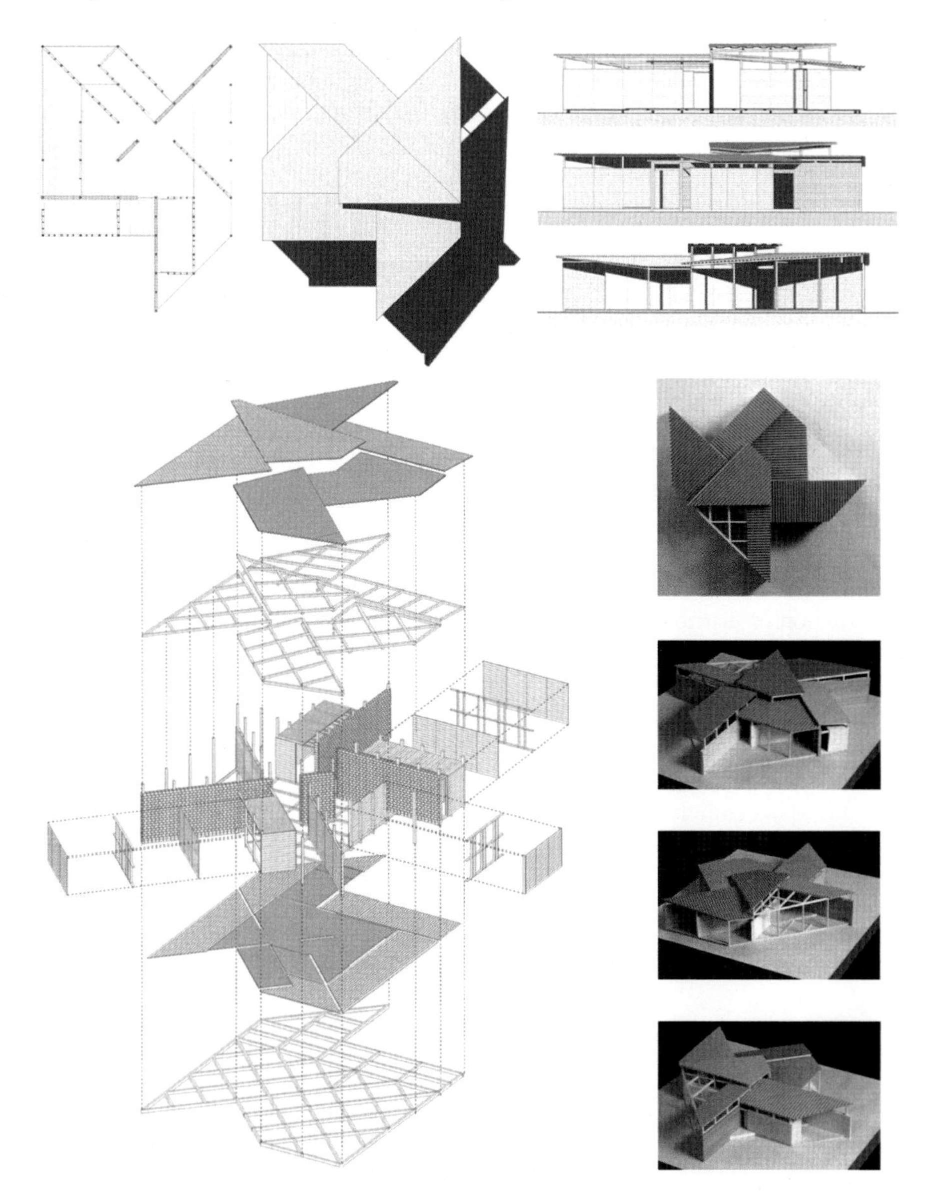

Fig. 4 Experimental design

possible scenarios that may be viewed as the abacus of possible solutions compatible with different contextual conditions. The underlying objective is to define a method of prototyping capable of assuming the degrees of versatility required by the contingency of the intended use, but which can also encapsulate an idea that envisages coherent typological and representational choices (Canella 1978). For this reason, we have chosen to work through a prototype rather than a defined design in an attempt to render as clear as possible the principles of generalisation as opposed to the specific contextual requirements to which a design tends to by its nature. On the other hand, the same envelope can potentially be designed to acquire solar energy to control shading and views, to insulate for natural and controlled ventilation, for natural and artificial lighting, to store water, etc. All essential requirements which, however, adhere to a principle of good practice, which it is believed, should be developed according to the specific design preferences that deviate from the generalisation conditions required here in order to make the adopted procedure transmissible. A unique construction system, therefore, is able to work for different intended uses, simply adapting its structural geometry to the function and modulating the space to the form's requirements. Not only that. The objective also relates to the design of typological versatility of the architectural artefact which, through an appropriate examination of distribution choices, is able to combine and overlap various activities compatible with those relating to welfare and health (education, education and training) in its daily operating cycle. And here, too, the generalisation aspects may subsequently, on a case by case basis, be adapted to the social and localisation needs of the context. In this way, from the standard minimum building structure capable of offering, for example, care and services to the most remote rural communities (doctor's visits, immunisations, pre-natal and post-natal care, nutrition centres, etc.), this proposal aims to investigate how this building principle could be generalised for more complex functions and spaces and for the associated activities and tasks necessary, in each case.

References

Canella G (1966) Il sistema teatrale a Milano. Dedalo, Bari, p 165

Canella G (1978) Assumere l'emergenza che non finisce. Hinterland 5–6:2–3

Ceccarini I (1989) Composizione modulare: grammatica della progettazione. U. Hoepli, Milan, pp 73–77; 157–162; 274–275

Chizzoniti D, Beggiora K, Cattani L, Moscatelli M (2014) Health post: a sustainable prototype for the third world. World Acad Sci Eng Technol Int J Civ Architectural Sci Eng 8(4):56–61

Falasca CC (2000) Architetture ad assetto variabile: modelli evolutivi per l'habitat provvisorio. Alinea, Florence, pp 13–29; 116

Kleczkowski BM, Montoya-Aguilar C, Nilsson NO (1985) Approaches to planning and design of health care facilities in developing areas, vol 5, no 91. Division of Strengthening of Health Services, World Health organization, Geneva, pp 31–33, 77–106

Neutra R (1948) Arquitetura social em paises de clima quente. Gerth Todtmann, São Paulo

Nicchia R (2011) Planning African rural towns. The case of Caia and Sena, Mozambique. University of Trento, Department of Civil, Environmental and Mechanical Engineering
Novi F (1994) Criteri e principi per la costruzione facilitata e l'autocostruzione con l'impiego di processi costruttivi, di strumenti e di tecnologie innovative: relazione finale del triennio di ricerca. BE-MA, Milan, pp 150–160; 176–184
Rama I (2013) Sviluppo Rurale e Assetto del Territorio in Contesti in Via di Sviluppo: un Progetto Attuativo del Plano de Uso da Terra in un Distretto Rurale del Mozambico. University of Trento, Department of Civil, Environmental and Mechanical Engineering
Semerani L (1978) Ricostruzione senza Rinascita. Hinterland 5–6:4–8
Staib G, Dörrhöfer A, Rosenthal M (2010) Atlante della progettazione modulare: elementi, sistemi, nuove tecnologie. UTET Technical Sciences, Turin, pp 25–30; 43–44

Energy and Built Environment

Introduction

Cinzia Talamo, Niccolò Aste, Corinna Rossi, Rajendra Singh Adhikari

Energy used in commercial and residential buildings accounts for a significant percentage of the total national energy consumption. It is estimated that 40% of the total electricity that is generated is used in buildings alone, consuming more energy than the transport and industry sectors. The building sector encompasses a diverse set of end-use activities, which have different energy use implications. The amount of energy used for cooling, heating and lighting is directly related to the building design, building materials, the occupants' needs and behavior, and the surrounding micro-climate.

The majority of modern buildings in Sub Saharan Africa (located in tropical climates) are replicas of buildings designed for the North European and North American regions (cold and temperate climates) and do not take into consideration the differences in climate. As a result, buildings are heavily reliant on artificial means for indoor comfort, i.e. cooling, heating and lighting. Thus, inefficient design, adoption of inadequate materials, combined with poor understanding of thermal comfort, passive building principles and energy conscious behavior, led to the current, dramatic energy waste.

Energy efficiency is one of the most important topics worldwide. In order to tackel this issue, it must be borne in mind that each region has its own specifics, which have to be considered before taking any decision or action. It is thus especially important to develop integrated design methodologies, and design tools able to deliver comprehensive information on energy efficiency and renewable energy technologies, to be used by designers, consumers as well as municipalities and governments. These design methodologies and tools are essential to achieve the energy retrofitting of existing building, to save energy, to design new energy efficient buildings, as well as to improve and design renewable energy technologies.

This section presents the results of research projects dealing with methodologies and tools designed to support the energy efficiency of buildings built in tropical African climates, ranging from efficient design of moden constructions, to low cost, context-adaptable to correctly store food and medicines.

Sustainable Building Design for Tropical Climates

Niccolò Aste, Federico M. Butera, Rajendra Singh Adhikari and Fabrizio Leonforte

Abstract The "Handbook-Sustainable Building Design for Tropical Climates" considers the impact of the construction sector on climate change, estimating that the building stock that will be built in Sub-Saharan Africa by 2050 will be three times greater than the current overall building stock of Europe. The purpose of the handbook is to offer an easy-to-use tool that provides general guidelines and basic information on the physics of buildings, together with all of the practical tools necessary for designing a sustainable-energy building in a tropical climate. The contents of the handbook were tested and evaluated by means of a series of lectures and training sessions—during the training courses on "Sustainable Integrated Design" held in different East African countries. These courses were attended by professionals (including engineers and architects), entrepreneurs, university teachers and postgraduate students. Further, based on the handbook, two "Massive Online Open Courses" (MOOCs) have been developed for the dissemination of the knowledge among various stakeholders.

Keywords Building energy efficiency · Sustainable buildings · Tropical climate · East African community · MOOCs

1 Introduction

Roughly half of the world's population lives in urban areas, and this proportion is projected to increase to 66% by 2050, adding 2.5 billion people to the urban population. Over the next two decades, nearly all the world's net population growth is expected to occur in urban areas, with about 1.4 million people added each week (GCEC 2014).

For such reason, the total building stock is expected to nearly double by 2050, at a rate of 5.5 billion square metres per year to almost 415 billion square metres in 2050, compared to the current global building stock of 223 billion square metres (GABC

N. Aste · F. M. Butera · R. S. Adhikari (✉) · F. Leonforte
Architecture, Built Environment and Construction Engineering—ABC Department, Politecnico di Milano, Milan, Italy
e-mail: rajendra.adhikari@polimi.it

N. Aste et al. (eds.), *Innovative Models for Sustainable Development in Emerging African Countries*, Research for Development, https://doi.org/10.1007/978-3-030-33323-2_4

2016). The current building stock is asymmetrically distributed among countries, as well as the estimated new building construction. Roughly 1 billion buildings are expected to be built in addition to the current 1 billion existing building, mainly located in OECD countries and China as shown in Fig. 1. The African continent, as well as India, still has less buildings than North America and Europe, but the total number of new buildings in the two regions is expected to overcome developed countries within 2050, with more than 50 billion m^2 and 30 billion m^2, respectively. In fact, by 2050, it is expected that the population of Africa will double, reaching 2.5 billion inhabitants. Thus, 700,000 new homes, 310,000 new schools and 85,000 new clinics are expected within 30 years, as well as additional facilities and infrastructures. Thus, the issues of the unsustainability of building sector growth and its impact on social, economic and environmental well-being have been addressed since the preliminary phase of the design process.

Currently, buildings represent an estimated 36% of global final energy consumption and 39% of the global energy-related carbon dioxide emissions [the latter comprised of 28% operational emissions and 11% from materials and construction (IEA 2018)]. Growing population and income levels in emerging economies and developing countries represent the main driver for building stock increases, implicating an estimated increase of energy demand in building equal to 50% by 2050 if no action is taken (GABC 2016). The trend is still increasing, even if important measures have been worldwide implemented in terms of energy efficiency, especially in OECD countries.

The goal of total decarbonization of building sector passes through the construction of new building with zero or almost zero-energy consumption from fossil fuels, i.e. zero carbon buildings, and the total renovation of the existing building to the same net zero carbon standards.

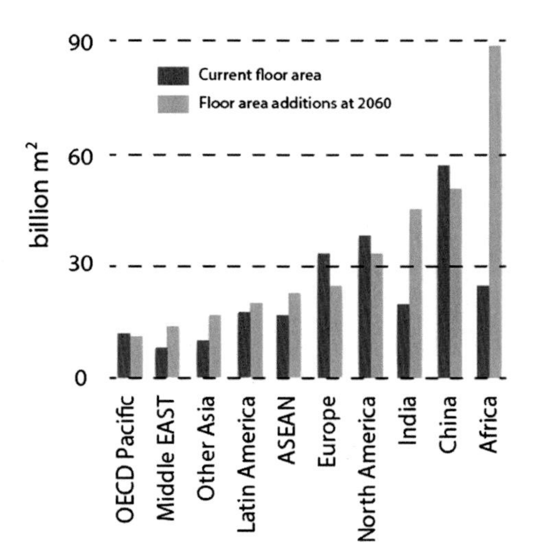

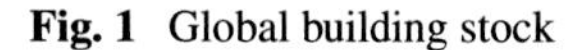

Fig. 1 Global building stock

Current renovation rates account for about 1% of existing building stock each year (GABC 2016). While to achieve 100% zero carbon goal by 2050, it is necessary to ensure a renovation rate higher than 3% (WGBC 2017).

CO_2 emissions resulting from material use in buildings represent almost one-third of building-related emissions, as concrete and steel manufacturing requires high amount of energy and implies large process emissions. As buildings become more efficient and the grid decarbonizes, embodied carbon increases in significance and will be responsible for nearly 50% of total carbon emissions of global new construction from 2020–2050. In this context, construction industry is required to radically change its manufacturing structure, in order to abate this increasing embodied energy.

In such a scenario, particularly, multifaceted and challenging are the role of the settlement design on minimizing energy consumption for cooling. This is a very critical issue, as energy consumption for cooling has been growing very steeply, and—according to the International Renewable Energy Agency (IRENA 2019)—will continue to grow, sustained by the climate change and the growing per capita income. For this reason, the issue of cooling must be specially highlighted.

Current models of human settlements are not designed to cope with the environmental challenges and are not sensitive to the rapid technological advancements from which our built environment could benefit. New real estate developments at the neighbourhood scale are producing a silent and uncontrolled Urban Revolution (Butera et al. 2018). The design of new building and urban developments is the key issue for coping with global warming and the quality of urban life, and, bearing in mind that urban design principles that apply to cities in tropical climates differ significantly from the principles that apply to cities in temperate climates, it is a burden shared by both developed and developing countries.

2 The Handbook of Sustainable Architecture in the East African Community

In such framework, the project "*Promoting Energy Efficiency in Buildings in East Africa*", an initiative of UN-Habitat in collaboration with the United Nations Environment Programme (UNEP), the Global Environment Facility (GEF), the governments of Kenya, Uganda, Tanzania, Rwanda and Burundi and Politecnico di Milano, was born. The first initiative of the project was aimed to the development of the "*Handbook of Sustainable Architecture in the East African Community*" written by Federico M. Butera as principal author (Butera et al. 2015). It offers an easy-to-use tool that provides general guidelines and basic information on the physics of buildings, together with all of the practical tools necessary for designing a sustainable-energy building in a tropical climate.

Fig. 2 Training session on handbook

The project team prepared an initial version of the handbook containing information on the various climates (building a climate map of the East African Community, defining six climate zones based on the impact of climate on the energy performance of buildings), the relations between the various climates and the energy response of buildings, their energy efficiency, design methods, the use of various technologies in the field of energy and zero-energy buildings and communities. The contents of the handbook were tested and evaluated by means of a series of lectures and training sessions—these too in provisional form—which were held in the cities of Dar es Salaam, Nairobi, Kampala and Kigali. The knowledge transfer was also extended to West Africa, through workshops in Douala, Cameroon.

The lectures were attended by professionals (including engineers and architects), entrepreneurs, university teachers and postgraduate students. The lessons included evaluation questionnaires on the handbook and training provided as well as discussion time sessions regarding the handbook and lessons (Fig. 2).

The evaluations and suggestions were used to improve the contents of the handbook and examine a number of issues in more detail, and to fine-tune the delivery of training in order to satisfy specific requests that emerged during the lessons. In addition to the above-mentioned subjects, the handbook (Fig. 3) contains five appendices: Principles of Building Physics, Principles of Thermal and Visual Comfort, Exercises, Case Studies and Integrated Design Applications. The handbook (Butera et al. 2015) is available as a free download on the UN-Habitat website (http://unhabitat.org/books/sustainable-building-design-for-tropical-climates/), allowing students, professionals and public administration technicians to consult, learn about and update their knowledge concerning the various themes connected with the building–energy relationship. Politecnico di Milano plays a particularly significant role in the potential for transferring knowledge that can have an impact on the national agendas—in this case the national energy agendas—of the East African Community countries, as

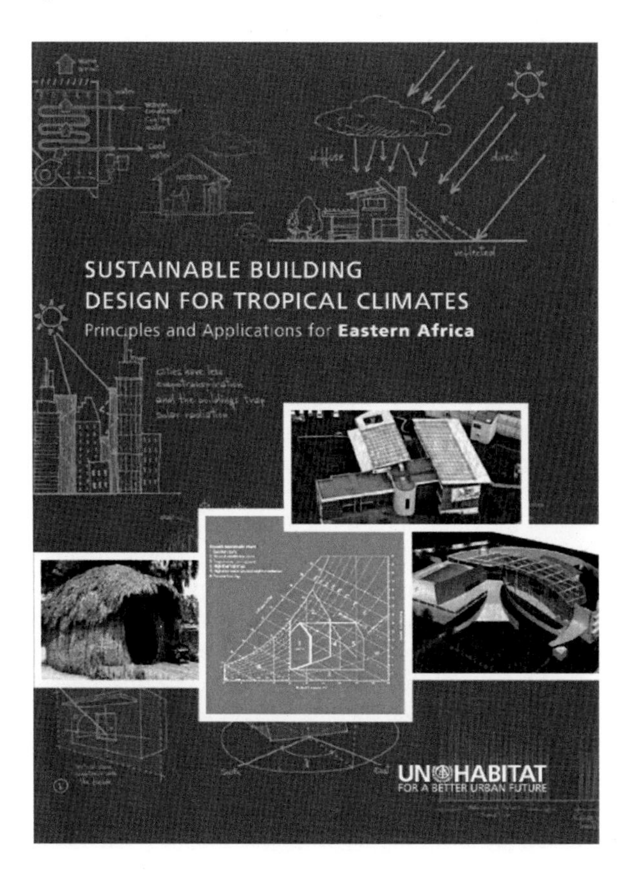

Fig. 3 Front cover of handbook

it is an internationally recognized institution with an extremely solid scientific reputation. The project has delivered two particularly important outcomes: firstly, the handbook has reorganized and disseminated a series of design-related information on tropical climates, providing dimensioning methods for the main parts of buildings in each of the different climate areas; secondly, it has produced a climate map based on building behaviour instead of the traditional ones based on vegetation.

According to the handbook, the following 30 strategies to consider in the design of sustainable building in the tropical climate can be outlined:

1. **Site analysis**: Assess the local context including the topography of the site. Collect data on temperature, relative humidity, wind's speed and direction, solar path and radiation.
2. **Building footprint**: The footprint of the building should ideally cover not more than 60% of the plot.
3. **Building orientation**: Design the long axis of the building to be along East–West to minimize direct solar radiation penetration in the building and reduce heat gain.

4. **Building shape**: Design it according to climatic zone. For hot-humid region, use narrow plans to maximize natural light, cross-ventilation and minimize heat gain. For hot-arid regions, use compact forms with courtyards to retain cold air in the building and minimize heat gain. Give preference to multi-storey building to increase density and maximize resources.
5. **Allocation of spaces within the building**: Services, e.g. toilets, starcases, lift, lobbies to be located on the East- and West-facing walls to act as buffer zones against heat gain but benefiting from daylighting.
6. **Opening**: Window sizing to be designed according to prevailing climatic conditions and placement preferably on North and South walls; window to wall ratio (WWR) should not exceed 40%. Glazing walls should be avoided, unless using special treated glass.
7. **Daylighting**: Design buildings according to climatic region, with openings on North and South walls, narrow plants to maximize daylighting, use light shelves in deep spaces. Window should be at least one-tenth of the floor area. The depth of the room should not exceed 2.5 times the height of the room.
8. **Solar protection**: Use sun shading devices, e.g. overhangs, vertical, horizontal shading elements, balconies, screens and vegetation to minimize heat gain.
9. **Natural ventilation**: Ensure that cross-ventilation is provided by the openings. Make use of roof vents and openings, thermal chimneys and clerestory windows. Make use of insulation materials under the roof sheet and design ventilated roofs.
10. **Cooling**: Integrate passive cooling systems by designing water bodies and features for evaporative cooling (just in hot and arid regions). Ensure that buildings using air condition appliances are well insulated to limit heat gains and reduce energy demand.
11. **Heating**: In highland region, enhance passive heat gain. Design passive solar heating strategies to ensure maximum sun penetration during cold season.
12. **Building envelope materials**: Always consider the carbon footprint content while choosing building materials. Give preferences to locally available building materials that are more appropriate with low energy content. Consider recyclable and reusable materials with low toxic emissions. Give preference to envelops (wall and roofs) with low U-value or low heat transmittance properties.
13. **External finishes**: Make use of light-coloured materials on external facades and roofs to reflect solar radiation in excess, while also incorporating green and living walls, vertical gardens provided with vegetation that grows on the facades.
14. **Renewable energy**: Integrate solar energy (thermal and electricity) such as photovoltaic and solar water heaters; wind energy, biogas and other available renewable energy systems into the building design.
15. **Water conservation and efficiency**: Design rainwater harvesting systems. Recycle grey water. Use water-efficient appliances and water-saving fixtures.
16. **Drainage**: Provide appropriate drainage technique to mitigate storm water runoff and facilitate replenishment of water table through rainwater infiltration.

17. **Sanitation**: In the absence of municipal sewage system, design on-site wastewater treatment facilities with production of biogas, compost and reused of water for irrigation.
18. **Solid waste management**: Design provisions for waste separation with on-site sorting facilities. Introduce innovative systems that encourage the 3R actions: reduce, recycle and reuse.
19. **Landscaping**: Design soft landscaping (greening site) with indigenous plants that require minimal irrigation and hard landscaping with paving materials that allow rainwater permeability. Limit paved areas around the building to reduce heat island effects.
20. **Energy-efficient appliances and energy demand management**: Incorporate energy-saving appliances in the building design. Make use of energy-saving bulbs, light level sensors, occupancy and motion sensors. Encourage behaviour change. Ensure that energy demand management principles are given top priorities by the building occupants.
21. **Well-balanced public spaces**: Fifty per cent of spaces should be allocated to streets, roads, public spaces, gardens and parks (30% far streets, 15% open space).
22. **Mixed land use**: Avoid zoning by combining economic, administrative and residential activities. This reduces the need to travel and ensures the use of public space.
23. **Mixed social structure**: Promote social integration and diversity. Encourage cosmopolitan values and the need to live together and avoid gated communities. Twenty to fifty per cent of residential space should be allocated to affordable housing.
24. **Adequate density and compact design**: High density neighbourhoods that are enough to trigger economies of scale and ensure livability.
25. **Connectivity**: Design street patterns and networks that connect the different parts of the city and ease the access to goods and services.
26. **Urban form matters**: Support mixed use, street life and walkability by designing compact blocks and buildings.
27. **Walkability**: Favour pedestrian mobility by emphasizing on walking distances, mixed use and public transport.
28. **Active mobility**: Street design should provide for pedestrians and cyclist lanes. Cycling extends reach of public transport.
29. **Promote the "shift"**: Encourage modal shift from energy-intensive modes, (cars) to walking, cycling and using public transport. Make cycling and walking safe and attractive.
30. **Promote vehicle efficiency**: Promote green transport by promoting the shift from fossil fuel-dependent vehicles to hybrid and electric cars.

3 Massive Online Open Courses (MOOCs)

In order to increase the dissemination of the knowledge developed in the handbook, two "Massive Online Open Courses" (MOOCs) have been developed. In detail, the "Massive Online Open Courses" (MOOCs) are online courses, implemented and issued with "open" logics, which generally involve a high number of participants, creating an equal community and which are distributed virtually on the global scale. These courses have spread rapidly since 2012 thanks to the launch of two international portals, Coursera and edX (http://www.coursera.org and http://www.edx.org), founded thanks to the encouragement of prestigious universities such as Harvard, Princeton, Stanford and MIT. All the initiatives connected with the Open Educational Resources, of which the MOOCs are an increasingly important pillar, playing a strategic role in redesigning the models of production and reproduction of knowledge and offering universities excellent occasions to test new forms of integration and share with the community on the local or global level.

The MOOCs—thanks to their being "open" and "massive"—are excellent channels through which to share ideas in local and global contexts and may therefore offer a significant contribution in this direction, but they also offer great networking opportunities. This potential develops in two dimensions; in the design phase: designing a MOOC allows the creation of flexible work groups that collect experts from different contexts in cultural, organizational and geographic terms, focusing on them with a short-term common goal and encouraging them to share the contents, opinions, etc.; in the execution phase: the global dimension of the MOOC facilitates the collaboration between the people interested in specific social themes fostering the creation of relationships on the global scale.

In this sense, the developed MOOCs called "*Sustainable building design for tropical climates: principles and guidelines for EAC*" and "*Sustainable building design for tropical climates: integrating design of buildings and technology systems*" aim to widely spread the basic skills for the design of sustainable buildings especially in tropical developing countries which, due to the exponential growth of their consumption, over coming years will prepare themselves to carry out a decisive role in the world's future energy scenario.

The courses are organized into different weeks and modules. At the end of each module, a quiz which checks the understanding of knowledge has been carried out (Fig. 4).

4 Conclusions

In conclusion, all the above-mentioned initiative aims to boost a technological change that implies also a cultural change. In order to face the issue of climate change, the culture of designers, architects, city planners, citizens, entrepreneurs and politicians has

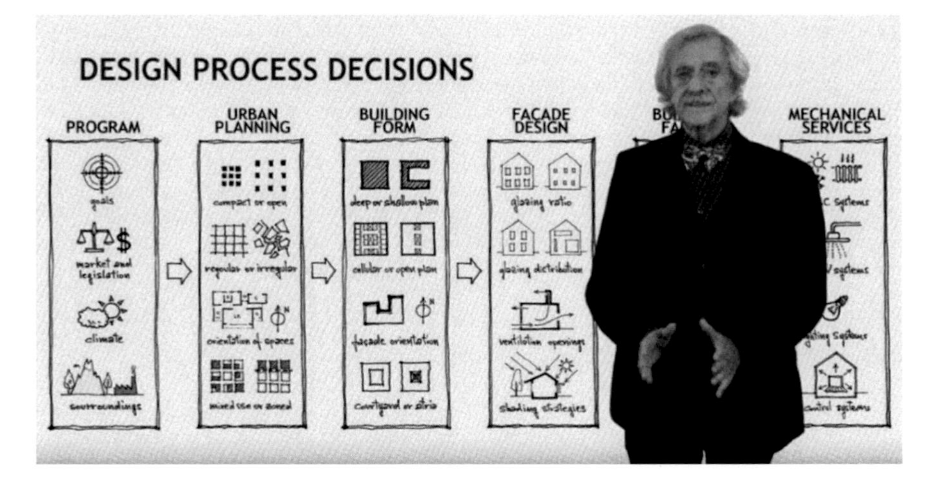

Fig. 4 Screenshot of MOOC developed based on handbook

to be changed. An unprecedented radical transformation of the methods of designing building and cities, especially in African countries, is required. The challenge for the new generations of designers is open.

References

Butera FM, Aste N, Adhikari RS (2015) Sustainable building design for tropical climates. UN-HABITAT, Nairobi, Kenya. ISBN 978-92-1-132644-4

Butera FM, Morello E, Pastore MC, Leonforte F (2018) Energy and resource efficient urban neighbourhood design principles for tropical countries. Practitioner's guidebook, UN-HABITAT, Nairobi, Kenya, HS Number: HS/058/18E

GABC (Global Alliance for Buildings and Construction) (2016) Towards zero-emission efficient and resilient buildings: global status report 2016

GCEC (The Global Commission on the Economy and Climate) (2014) New climate economy, better growth better climate. www.newclimateeconomy.report

IEA (International Energy Agency) (2018) Towards a zero-emission, efficient, and resilient buildings and construction sector

IRENA (International Renewable Energy Agency) (2019) Global energy transformation. A roadmap to 2050

WGBC (World Green Building Council) (2017) From thousands to billions. Coordinated action towards 100% net zero carbon buildings by 2050

Development of an Interactive Building Energy Design Software Tool

Niccolò Aste, Rajendra Singh Adhikari, Claudio Del Pero and Fabrizio Leonforte

Abstract Energy efficiency is one of the most important topics worldwide, nevertheless, each region has its own specifics, which have to be considered. This is especially important for online tools that deliver comprehensive information on energy efficiency and renewable energy for average consumer as well as for municipalities and governments. These tools are essential for energy retrofit of existing building for energy saving as well as designing of new energy-efficient buildings and also for designing and sizing renewable energy technologies such as solar home system. A dynamic interactive building energy design software tool has been developed especially tailored taking care about main features required for Sub-Saharan African context, under a joint agreement (research project) between Politecnico di Milano and UN-Habitat. In this chapter, the results of testing of the software tool for building geometry have been presented.

Keywords Building energy efficiency · Design tool · BESTenergy · Testing

1 Introduction

Energy used in commercial and residential buildings accounts for a significant percentage of the total national energy consumption. It is estimated that 40% of the total electricity generated in the region is used in buildings alone, consuming more energy than the transport and industry sectors. The building sector encompasses a diverse set of end-use activities, which have different energy use implications. The amount of energy used for cooling, heating and lighting is directly related to the building design, building materials, the occupants' needs and behavior and the surrounding microclimate. Majority of modern buildings in Sub-Saharan Africa (mainly tropical climates) are replicas of buildings designed for the western world (cold and temperate climates) and do not take into consideration the differences in climate. As a result, buildings are heavily reliant on artificial means for indoor comfort, i.e., cooling,

N. Aste · R. S. Adhikari (✉) · C. Del Pero · F. Leonforte
Architecture, Built Environment and Construction Engineering—ABC Department, Politecnico di Milano, Milan, Italy
e-mail: rajendra.adhikari@polimi.it

N. Aste et al. (eds.), *Innovative Models for Sustainable Development in Emerging African Countries*, Research for Development, https://doi.org/10.1007/978-3-030-33323-2_5

heating and lighting. The problem is that inefficient design and construction using inadequate materials, combined with the poor understanding of thermal comfort, passive building principles and energy conscious behavior, have led to tremendous energy wastage.

Therefore, energy efficiency is one of the most important topics worldwide, nevertheless, each region has its own specifics, which have to be considered. This is especially important for online tools that deliver comprehensive information on energy efficiency and renewable energy for average consumer as well as for municipalities and governments. These tools are essential for energy retrofit of existing building for energy saving as well as designing of new energy-efficient buildings and also for designing and sizing renewable energy technologies such as solar home system. Addressing these issues, the UN-Habitat programme "Promoting Urban Energy for climate change in developing countries" has the objective to mainstream energy efficiency and renewable energy measures in the delivery of sustainable and friendly built environment.

2 Interactive Building Energy Design Software Tool

Nowadays' need to meet sustainability requirements, limit environmental impacts and reduce operating costs imposes a structural modification of the architectural design process (Fig. 1), calling for an integrated approach in building energy performance analysis and planning, from the very preliminary design phase. In this framework, an accurate estimation of buildings' energy consumptions, related to the different uses and needs, becomes an essential design and evaluation tool.

At present, the most widespread methods to evaluate building energy performance rely on simplified steady-state calculations based on average reference data. Such methods are quite obsolete and inaccurate, as they do not properly account for the complexity of phenomena impacting energy performance (thermal inertia, temporal

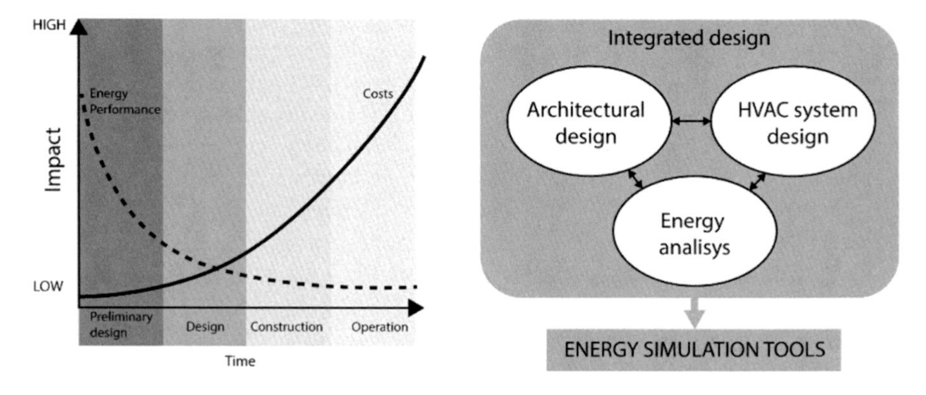

Fig. 1 Impact of energy expertise integration and integrated design approach

fluctuations, peaks and dips of boundary conditions) and prove to be particularly inadequate in the case of tropical climates and cooling-dominated buildings. In these scenarios, characterized by dynamic paths affecting building thermal behavior and energy performance, designers need more accurate, hourly based methods. These methods, which take into consideration the actual building's thermal response (heat absorption and release induced by the variation of outdoor conditions at different times of the day) are named dynamic state calculations.

Dynamic state calculations are processed by simulation engines and, due to their level of sophistication, are usually difficult to run, especially when they are not provided with a graphical user interface.

In order to enforce an integrated design approach inclusive of energy performance aspects, adequate tools must be developed, able both to correctly encompass the complex interactions between energy flows and design choices and to be user-friendly.

To overcome the above difficulties, Politecnico di Milano in collaboration with UN-Habitat developed an open-source, free-redistributable, intuitive and powerful building energy simulation software, especially tailored taking care about main features required in the African context.

This energy tool provides the following main characteristics:

- Reliable dynamic building energy simulation that is the only method which ensures a correct evaluation in cooling-dominated buildings and in tropical climates, where dynamic paths heavily affect buildings' thermal behavior and energy performance, in which designers need accurate, hourly based methods, considering the actual building thermal response and its heat absorption and release due to the variation of the outdoor conditions throughout the day. This is performed by EnergyPlus simulation engine, which is the state of the art in this item.
- The hardness in compiling a building energy model is avoided by an intuitive interface based on SketchUp, coupling, in this way, the easiness in geometric modeling of the modeling tool with the reliability and accuracy of EnergyPlus. In fact, users can build an energy model by using SketchUp drawing tools and dedicated interfaces, so that the plugin creates the input data file to be processed by the simulation engine with few easy steps. Also, some macros are developed to generate geometries and to automate modeling process.

In such respect, the so-called BESTenergy was born. A conceptual scheme of the software is shown in Fig. 2. BESTenergy includes the state-of-the-art knowledge in building energy estimation and at the same time supports architectural choices. Through this platform, users can keep track in an interactive way of all the consequences emerging from building energy behavior in relation to design choices by considering the overall energy performance—simultaneously including heating, cooling and lighting demands.

More in detail, BESTenergy takes advantage of a SketchUp plugin to run dynamic state building energy simulations using the EnergyPlus (Crawley et al. 2001) simulation engine, thus coupling the handiness of SketchUp geometric modeling with the reliability and accuracy of EnergyPlus (Crawley et al. 2005).

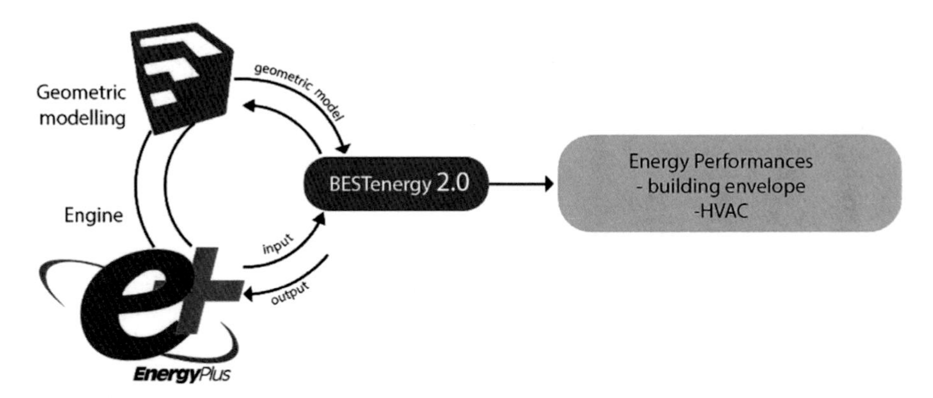

Fig. 2 BESTenergy software conceptual scheme

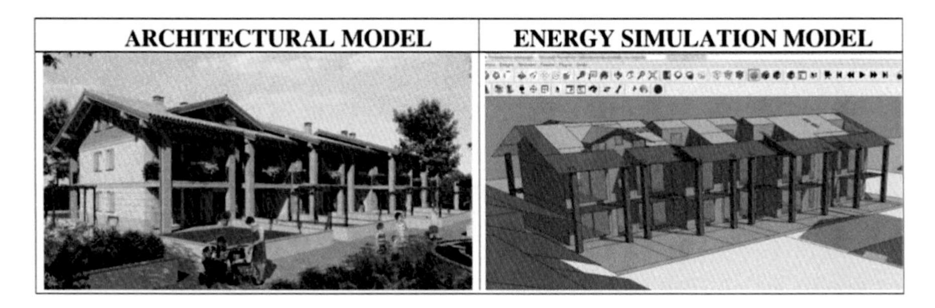

Fig. 3 Comparison between the architectural model and the simulation model in BESTenergy

Thanks to BESTenergy, users can build an energy model exploiting SketchUp drawing tools and dedicated interfaces, while the plugin tool creates an input data file to be processed by the simulation engine. A comparison between a real architectural model and energy simulation model has been shown in Fig. 3.

3 BESTenergy Simulation Platform

BESTenergy allows to comprehensively optimize energy resources in the building design phase by looking for the best compromise among different energy services. It also proves to be particularly suitable for evaluating real performances of an already designed (or in progress) building, providing accurate estimates of its features under operating conditions.

In particular, BESTenergy can evaluate

- heating, cooling and lighting energy demands on the same model and their interaction, letting the user to interactively optimize overall energy behavior;

- effects of movable shading devices;
- thermal comfort levels both by using adaptive comfort models and more traditional ones;
- natural ventilation effects, coupling an energy model with a ventilation model, which is usually split into different evaluation tools;
- local evaluations, as behaviors of each building surface, as its temperature, its amount of heat gain and so on;
- daylighting and simple lighting systems;
- hourly power demand;
- HVAC systems sizing;
- HVAC system hourly operation and response.

Because of its relevance in warm environments (African context), a special effort was dedicated to developing interfaces to model natural ventilation effects: The EnergyPlus airflow network mass transfer model included in BESTenergy enables users to calculate at each time step the actual ventilation flow taking into account the openings' geometry, their relative positions and opening time intervals. These evaluations are directly coupled to the overall building thermal model, and therefore, they interactively affect simulation outputs, avoiding a separate evaluation based on other specific tools.

During its development stages, BESTenergy was tested in university courses, professional trainings and in the context of research studies and professional consultations (Fig. 4), allowing to gather a sample of users characterized by a quite broad range of knowledge backgrounds. As a result, significant on-field feedbacks emerged that proved useful in understanding how to improve the platform's friendliness and functionalities.

Fig. 4 Students and tutors dealing with BESTenergy during a workshop

4 Testing Building Geometry Generated by the BESTenergy Tool

New tools must be tested to verify the validity of results of their use and to identify possible peculiarities and qualifications of how they generate those results. To verify building geometry modeled by BESTenergy, the research team selected the design of a three-story hypothetical office building (Fig. 5) that features non-trivial geometry configuration and has been previously repetitively used in the testing of other new simulation tools and utilities. It is known as the BLIS Test Building, developed jointly by the Interoperability Team at the Environmental Energy Technologies Division of the Lawrence Berkeley National Laboratory and Digital Alchemy.

The total net test building area is 1639.82 m^2 and gross roof area is 574.63 m^2 with an expected building occupancy between 90 and 100 people.

The wireframe of the building geometry model generated by BESTenergy is shown (Fig. 6); since it was drawn with SketchUp, it is a "center-line" model (walls' and slabs' thickness is not shown) which defines area and volume values that are larger than they really are. For comparison with the geometry generated by BESTenergy, the wireframe model of the test building geometry that is a full 3D CAD geometry model transformed for use in EnergyPlus is also shown (right).

Testing of the BESTenergy tool involved four steps: (1) drawing building geometry of the BLIS test building with SketchUp and applying BESTenergy as the interface to create a geometry compatible with EnergyPlus input requirements; (2) recording EnergyPlus geometry output data resulting from the simulation with BESTenergy; (3) executing the EnergyPlus Input Definition File (IDF), generated by the BESTenergy run, in independent "stand-alone" EnergyPlus runs; (4) comparing the reported detailed building geometry information to the original geometry information for the

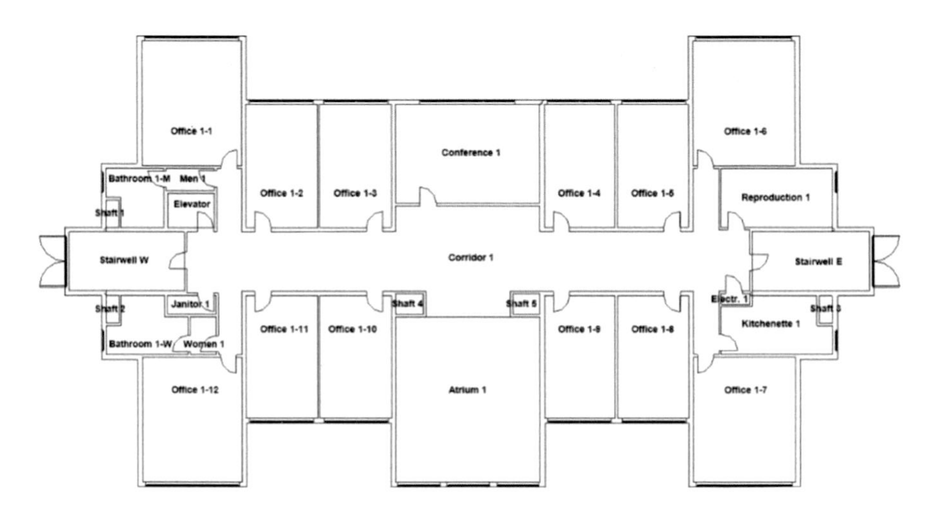

Fig. 5 Typical floor of the test building

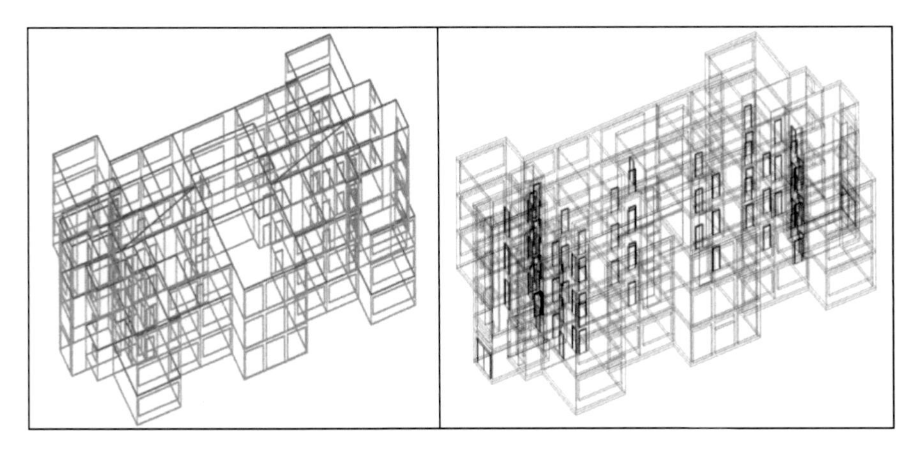

Fig. 6 Wireframe image of the test building model geometry generated with BESTenergy and imported into EnergyPlus (left) and wireframe image of the original test building geometry model generated in Revit 2018, then transformed and imported into EnergyPlus (right)

BLIS building that had been generated by Revit and subsequently semiautomatically transformed and run by EnergyPlus.

All EnergyPlus runs related to the testing of the BESTenergy tool and used in the comparison were executed with the EnergyPlus version 8.9 simulation engine (which currently is the latest EnergyPlus release). The original geometry of the BLIS test building was first defined in ArchiCAD (Graphisoft 2018)—and subsequently also in Revit (Autodesk 2018)—and then extended with space boundary definitions automatically calculated by SBT and verified with the Solibri Model Checker (Solibri 2018). Modelers involved in the testing and verification of the BESTenergy tool are experienced in building geometry modelers with a long record of involvement with BEP simulation.

First of all, the net floor areas for each space/zone in the test building (called "BESTenergy Area"), calculated by EnergyPlus using the geometry model generated by the BESTenergy tool, have been compared with net floor areas of the original geometry model generated by Revit. After that, a similar evaluation was carried out on the volume. A sample of the obtained data is reported in Table 1.

After that the "Difference in Area" that represents difference between the same space/zone area values calculated from the BESTenergy geometry model and the original BLIS building geometry model has been evaluated. In each case, the difference is expressed as percentage of the smaller value in the pair. Calculated differences vary widely: The smallest difference is 3.06%; the largest is 18.25%. This difference is caused by the center-line model and cannot be standardized, as it depends on the actual thickness of every building element (wall or slab) that is represented in that center-line geometry model.

In fact, EnergyPlus geometry model is a surface geometry model. It accepts building surface information as defined in its input file and calculates itself all other non-surface information it may need from available surface data (such as space/zone

Table 1 Quantitative comparison of building geometries in test

Space/zone name	On floor(s)	BESTenergy area (m^2)	Test building area (m^2)	Difference in area (%)	BESTenergy volume (m^3)	Test building volume (m^3)
OFFICE 1-1	1	28.99	28.07	**3.28**	92.76	85.33
OFFICE 1-2	1	20.67	19.35	**6.82**	66.16	58.82
OFFICE 1-3	1	20.62	19.29	**6.89**	65.97	58.64
OFFICE 1-4	1	20.67	19.29	**7.15**	66.16	58.64
OFFICE 1-5	1	20.61	19.35	**6.51**	65.97	58.82
OFFICE 1-6	1	28.99	28.07	**3.28**	92.76	85.33
ELEVATOR	1, 2, 3, 4	4.25	3.62	**17.40**	52.33	45.90
CONFERENCE 1	1	32.90	31.48	**4.51**	105.28	95.70
REPRODUCTION 1	1	15.60	14.55	**7.22**	48.67	44.23
MEN 1	1	2.77	2.45	**13.06**	8.87	7.45
BATHROOM 1-M	1	7.60	6.71	**13.26**	24.31	20.40
SHAFT 1	1, 2, 3	0.87	0.79	**10.13**	8.18	7.42
OFFICE 1-7	1	28.99	28.07	**3.28**	92.76	85.33
OFFICE 1-8	1	20.61	19.35	**6.51**	65.97	58.82
OFFICE 1-9	1	20.67	19.29	**7.15**	66.16	58.64
OFFICE 1-10	1	20.62	19.29	**6.89**	65.97	58.64
OFFICE 1-11	1	20.67	19.35	**6.82**	66.16	58.82
OFFICE 1-12	1	28.99	28.07	**3.28**	92.76	85.33
CORRIDOR 1	1	107.80	101.45	**6.26**	344.95	308.41
ATRIUM 1	1	55.30	52.83	**4.68**	176.96	169.06
KITCHENETTE 1	1	13.60	12.75	**6.67**	43.53	38.76
WOMEN 1	1	2.72	2.17	**25.35**	8.71	6.60
BATHROOM 1-W	1	9.31	8.82	**5.56**	29.8	26.81
JANITOR 1	1	2.42	2.07	**16.91**	7.73	6.29
ELECTRICAL CLOSET 1	1	0.70	0.67	**4.48**	2.25	2.04
SHAFT 1	1, 2, 3	0.87	0.79	**10.13**	8.36	7.42
SHAFT 2	1, 2, 3	0.87	0.79	**10.13**	8.18	7.42
SHAFT 3	1, 2, 3	0.87	0.79	**10.13**	8.18	7.42
SHAFT 4	1, 2, 3	1.49	1.26	**18.25**	13.96	11.83
SHAFT 5	1, 2, 3	1.42	1.26	**18.25**	13.29	11.83
Average conditioned	3	26.92	25.50	**5.57**	∑ (areas)/no. of conditioned zones	
Average ventilated	3	8.46	7.77	**8.89**	∑ (areas)/no. of ventilated zones	

Conditioned
Ventilated
Unconditioned
Area difference

volumes), even if such information is available in the input file. Because of that, any comparison of volume data does not add new information, as any volume value is a function of the associated floor area value and is recalculated by EnergyPlus.

Definition of building geometry with "center-line" tools inevitably causes quantitative errors if compared with the real 3D geometry of the given building. Correct geometry can be manually calculated from "center-line" geometry, as illustrated Fig. 7.

Yet, "center-line" CAD tools are popular. They are easier to learn and use than object-oriented model-based CAD tools, require less effort and modeling time, and the modeling software is usually much less expensive. SketchUp is a good example: It is the recommended geometry definition software for OpenStudio applications; it is particularly often used by novices as the graphical user interface (GUI) to building energy performance (BEP) simulation with EnergyPlus that directly defines building

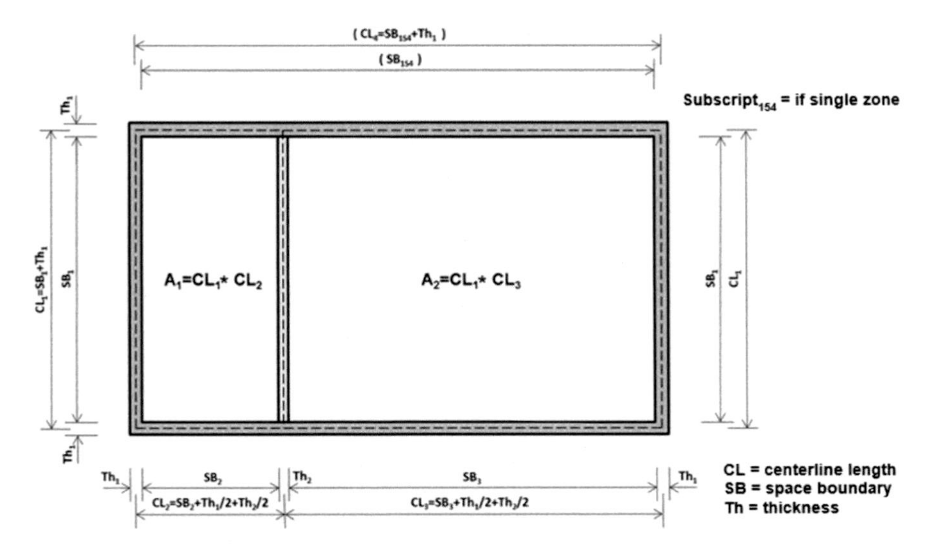

Fig. 7 Example of calculating correct floor area from "center-line" geometry definition (Bazjanac et al. 2016)

geometry for the simulation even though the defined geometry is only an approximation.

The critical question then is what is acceptable engineering error in BEP modeling and simulation? The International Standards Organization (ISO 2018) and other standardization bodies offer no help in defining acceptable error in BEP modeling and simulation. Search of the Internet reveals a lively discussion of the subject within the buildings industry and an apparent consensus among its participants: The acceptability of a percent error depends on the application. Buildings industry consensus for acceptable errors in numerical results appears to be 5% or less.

Perhaps the critical question above should be refined: What is acceptable accuracy, precision or error that makes a given building geometry model still useful? If one must use data showing high error values, one should (1) understand the cause(s) of error(s); (2) assess the significance and impact of error(s) in question and (3) if available/possible, compare calculations/predictions to measured values, then calibrate the model accordingly.

In conclusion, BESTenergy can be considered a reliable tool to use for energy evaluation, especially when overestimation is not harmful and results are informative and useful. Overall, "center-line" error depends on the sum of ratios of all object thicknesses, not on the lengths or heights of the corresponding walls and slabs. It often significantly exceeds 5%. Building geometry defined by SketchUp, and consequently, used by the BESTenergy tool, is an approximation of the true geometry of the building being modeled; using it in EnergyPlus will overestimate building's thermal loads and will likely result in oversized HVAC systems selected for the building. Yet, by using the BESTenergy tool as a GUI to prepare input for EnergyPlus, modeling building geometry is significantly less challenging and less time consuming than developing full CAD-type geometry models.

The differences between BESTenergy and the original BLIS building geometry models can be directly attributed to the difference in the types of geometry modeling involved. For the test building geometry used in the BESTenergy generated geometry model evaluation, the differences exceed the apparent commonly acceptable industry error tolerance of 5.00%. The "center-line" modeling approach leaves out the thickness of the modeled objects (walls or slabs); this inevitably results in the definition of larger areas and volumes than those from modeled by object-oriented CAD.

Modeling building geometry for use in EnergyPlus requires a certain level of precision. The quality of modeling and BEP simulation depends directly on the modeler's ability and provided effort. Novice modelers and modelers with little previous exposure to modeling knowledge will find BESTenergy to be a most useful tool which quickly advances their modeling knowledge and skills.

Value and utility of BEP modeling and simulation are often directly proportional to the skill of the modeler. Successful BEP modeling and simulation require several skills: (1) knowledge and understanding of BEP simulation; (2) working knowledge of model-based CAD and other simulation-related software tools; (3) good understanding of building physics; (4) understanding of building systems and HVAC and (5) ability to effectively engage in large database analysis. Successful modeling is a collaborative process involving, besides the modeler(s), the entire building design team, project management and consultants, as well as relevant government officials. The buildings industry would greatly benefit from a larger number of skilled modelers than available in the currents pool. Modeling is not a skill acquired at birth; it is a skill that is learned and requires experience. For that, future modelers need access to properly designed, funded and executed educational programs, as well as timely opportunities and patience to practice newly acquired and developed skills.

References

Autodesk (2018) Revit Architecture 2018. https://www.autodesk.com/products/revit/architecture

Bazjanac V, Maile T, Nytsch-Geusen C (2016) Generation of building geometry for energy performance simulation using Modelica. In: Proceedings of the CESBP/BauSIM 2016 conference, Dresden, September 2016

Crawley DB, Lawrie LK, Winkelmann FC, Pedersen CO (2001) EnergyPlus: a new-generation building energy simulation program. In: Proceedings of forum 2001: solar energy: the power to choose, Washington, DC, April 2001. ASES, Boulder, Colorado

Crawley DB, Hand JW, Kummert M, Griffith BT (2005) Contrasting the capabilities of building energy performance simulation programs. In: Proceedings of building simulation 2005, Montreal

Graphisoft (2018) ArchiCAD 2022. https://www.graphisoft.com/archicad/

Solibri (2018) Solibri model checker. https://www.solibri.com/howit-works

SPARK—Solar Photovoltaic Adaptable Refrigeration Kit

Claudio Del Pero, Maddalena Buffoli, Luigi Piegari, Marta Dell'Ovo and Maria P. Vettori

Abstract In developing countries, especially in particularly critical areas such as tropical Africa, the state of health of the population is strongly influenced by infectious risk factors. High temperatures and humidity levels, in fact, contribute to the formation and proliferation of viruses and bacteria capable of spreading rapidly. For these reasons, the need to find low-cost and context-adaptable strategies capable of ensuring the correct preservation of food and medicine in areas where the electricity supply is not reliable is a priority for improving health and social conditions in developing countries. In such a context, the aim of this study was to develop a modular kit designed for the self-construction of a refrigeration system which is economically competitive with other products already developed for the same purpose. The refrigeration system is powered by solar photovoltaic energy and can be easily assembled in the required location of application, with particular reference to the context represented by Africa's tropical belt. The kit was designed and tested at the Politecnico di Milano University and the final prototype version was built in Cameroon.

Keywords Solar refrigeration · Photovoltaic (PV) refrigerator · Food preservation · Self-construction · Low-cost refrigerators

1 Introduction and Context Characterization

The unfair access to fundamental necessities, such as clean water, uncontaminated food, drugs, etc., strongly influences the well-being of a Country and is a cause of death (WHO 2018; Haver et al. 2013). Particularly, the most affected countries are those belonging to the tropical–equatorial belt of Africa and characterized by a low Human Development Index (Del Pero et al. 2015).

C. Del Pero (✉) · M. Buffoli · M. Dell'Ovo · M. P. Vettori
Architecture, Built Environment and Construction Engineering—ABC Department, Politecnico di Milano, Milan, Italy
e-mail: claudio.delpero@polimi.it

L. Piegari
Department of Electronics, Information and Bioengineering—DEIB, Politecnico di Milano, Milan, Italy

N. Aste et al. (eds.), *Innovative Models for Sustainable Development in Emerging African Countries*, Research for Development, https://doi.org/10.1007/978-3-030-33323-2_6

Their economic activity is mainly based on the primary sector, specifically on agriculture, but few preservation practices are adopted; in fact, more than the 30% of the production is discarded due to incorrect food storing practices (Haver et al. 2013). Among the main reasons, there is a lack of appropriate technology and a very limited access to energy, especially electricity, usually concentrated in major cities and restricted to less than 10% of the inhabitants. In this context, in fact, rural areas' settlements are the most affected by this problem (Rebecchi et al. 2016) since they are typically not served by the grid or with a very limited access only for a few hours per day (IMECHE 2014; Tassou et al. 2010). To overcome this issue, several stand-alone solutions for food preservation have been developed providing different innovative alternatives from which it is possible to choose the most suitable for the application context, such as those that are based on sorption refrigeration, using either fossil fuels or solar thermal collectors as a heat source (Somerton et al. 2009; Metcalf et al. 2011; N'tsoukpoe et al. 2014; Santori et al. 2014; Yildiz 2016). More recently, DC (Direct Current) PV (Photovoltaics) refrigeration has become a very interesting and affordable solution, mainly thanks to the reduced manufacturing cost of photovoltaic modules (Nemet 2006; Feldman et al. 2012).

Considering the importance of the topic and the increasing market demand, several commercial solutions are currently available (Ewert et al. 1998; Ewert and Bergeron III 2000; Kim and Ferreira 2008), but most of the products are characterized by high cost and certain limits:

- the use of batteries for energy storage in case of electric refrigeration;
- the implication of complex technical solutions given by the use of the absorption cycle coupled with solar thermal collectors;
- the manufacturing of the product in developed countries and the shipping as a ready-to-use system.

Given these premises, the aim of the present work is to describe a solar photovoltaic adaptable refrigeration kit (SPARK); it is an assembly kit of a stand-alone DC solar refrigerator powered by PV energy, suitable for food conservation in remote areas of countries in a state of protracted crisis (Critoph and Thompson 2002; Freni et al. 2008; Kanade et al. 2015). The project is the result of a study funded by the Politecnico di Milano (Polisocial Award competition) and developed in collaboration with African Center for Renewable Energies and Sustainable Technologies (ACREST), an NGO located in Cameroon, where the solution has been tested. In fact, the most suitable area selected for the application of the system is the tropical–equatorial belt of Africa (Fig. 1).

The work is organized in three sections: the methodology, aimed at presenting the phases of development; the results, to describe the product and the validation stage; and the conclusions.

Fig. 1 Application context of the project

2 Methodology

The proposed methodology is articulated in the following different phases.

PHASE 1: analysis of the state of the art and characterization of the context.

This phase examines in depth the framework of rural communities in tropical African countries, with a specific focus on the territorial/climatic/social/economic contexts and on the availability of local materials/technology; the aim is to create, in the following phases, a product that is truly useful, efficient and, above all, economically and technically feasible on a large scale. In particular, the focus of the research was on the village of Bangang in Cameroon (80,000 inhabitants), which was chosen as a representative application context.

PHASE 2: project preparation.

The information and analyses carried out are systematized according to the following macro-activities:

- optimization of commercially available and/or re-functionalized technical components that can be used in the specific context;
- development of a management and control system to ensure the correct functioning of all components;
- definition of sizing and self-construction criteria for the thermal envelope and energy storage system,
- system design + drafting of the guidelines for the final assembly of the kit.

In this phase, the prototype of the refrigeration kit was also assembled.

PHASE 3: project experimentation and validation.

The product was first experimented in laboratory conditions at the Politecnico di Milano and then was validated in Cameroon through an assisted self-construction activity. This phase also included the dissemination of the results with the support of local NGOs.

3 Results

After the consistent analysis of the state of the art aimed also at investigating existing practices and technologies (Del Pero et al. 2015), (Phase 1), the solar refrigerator powered by PV energy was designed in detail (Phase 2). With regards to its details, the system is characterized by the use of local materials and manpower, moreover it is modular, battery-free, and can be self-constructed on-site. The following features were defined as main requirements for the development phase:

- low-cost, in order to ensure the highest possible number of users guaranteed by the reduction in the shipping and manufacturing phase;
- high reliability, in fact the system is designed to be composed of reliable and durable components which can be easily repaired or substituted by the locals;
- high hygienic performance, guaranteed by the waterproof structure of the refrigerator, by a proper drain of the condensate and by specific guidelines on food preservation (Capolongo et al. 2012);
- modular structure: the basic module of the system (with a capacity of 250 L) can be easily coupled with other modules in order to create refrigerated volumes of different sizes;
- high thermal storage, by ensuring full operation without solar power for at least 72 h (Del Pero et al. 2018);
- reduced environmental impact, in order to minimize any kind of pollution during or at the end of its working life, in the proposed system all of the components can be re-used or recycled;
- active participation of local people in the manufacturing phase, to promote the economic benefits and improve technical knowledge (Capolongo et al. 2011). In detail, the system can be assembled without the participation of skilled workers

with specific expertise or technical equipment, in fact, the import of ready-to-use products from developed countries can reduce the level of local technical knowledge.

System Description

Based on the above-listed requirements, the precise definition of the technological configuration of the kit was carried out. In particular, the PV refrigerator is composed of the following main elements: A PV module, a DC compressor with a forced-air-cooled condenser, a roll-bond evaporator, a thermally-insulated envelope made with wood and thermal insulation, an ice storage, an expansion valve, and the control system. More in detail, the PV module generates DC electrical energy which powers the small-size DC compressor. The latter activates a refrigeration cycle, with a forced-air condenser and an evaporator placed inside the volume to be refrigerated. The cooling energy is used to freeze a mass of water (thermal storage) that allows the internal temperature to be maintained acceptably low both during the night and during any period with poor irradiance (full operation without solar power equal to 2–3 days) (Fig. 2).

The DC-powered compressor with variable input voltage was chosen with the specific purpose to avoid the use of inverters and batteries: in fact, the electronic unit of the compressor allows for direct coupling with PV modules. A pre-assembled product, typically used in the nautical field, was selected; it represents a product already well-tested under severe conditions. An air-forced condenser was preferred to a static one, because it ensures a good heat exchange even under conditions of high temperature and poor ventilation.

For the PV generator, a polycrystalline silicon PV module with high efficiency was selected. It is made of 60 polycrystalline silicon cells (156 mm × 156 mm) and is designed to work in the most difficult environmental and operating conditions.

For the evaporator side, a one-side-flat roll-bond heat exchanger was identified as the best choice, in order to ensure the maximum heat transfer surface and consequently improve the heat exchange. The component is made in aluminum and is suitable to work in direct contact with water (dimensions: 585 mm × 345 mm). It is in fact directly immersed in an aluminum or plastic tank placed inside the refrigerated volume and filled with water, to act as a thermal storage system.

The main function of the control system is to manage energy uses and to maintain the desired temperature level inside the insulated case, indicating also when it is possible to introduce food/beverages to be refrigerated in the system. Considering the specific configuration, the control system was specifically developed within the research project and supplied as a component within the assembly kit.

- The envelope of the refrigerators is manufactured on-site with local materials and skills and it is composed of:
- internal layer: local wood (0.9 mm) covered by a waterproof foil;
- thermal insulation: bamboo's core (150 mm);
- external layer: bamboo (150 mm).

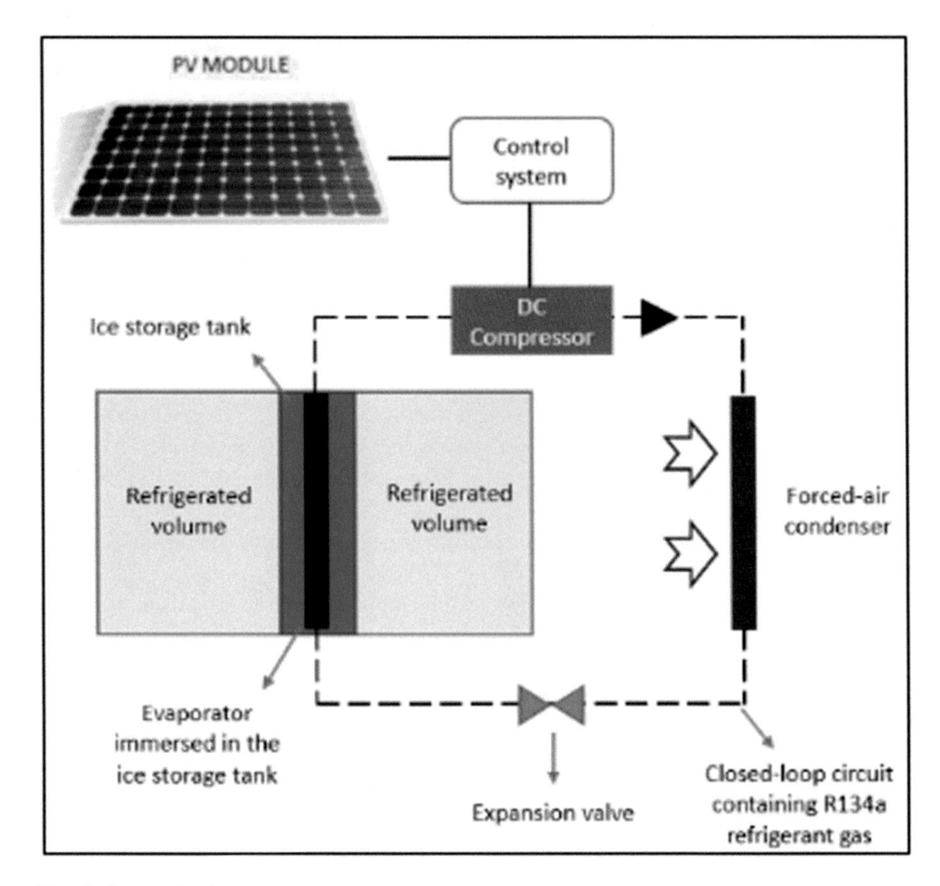

Fig. 2 System basic structure

The internal waterproof layer can be made of liquid-state plastic which can be applied like a varnish or by a more traditional plastic foil, while the bottom of the box is covered by an aluminum foil (1 mm). The box is lifted off the ground to stay dry and away from possible attacks by small animals. The upper side of the box is protected by a pressure cap and the leak-tightness is guaranteed by its own weight. According to the availability of local resources, it is possible to use other materials or construction techniques. The thermal transmittance of the described envelope is 0.3 W/m^{2}K (Fig. 3).

SPARK was first tested at the laboratory of the Politecnico di Milano in order to guarantee its efficiency and to stress its limits, resistance, and robustness (Del Pero et al. 2016).

In addition, to assess the energy performance of the system, a simplified energy model was developed. The model estimates the mean air temperature inside the refrigerator and the temperature of the water in the storage container, on the basis of the boundary conditions, assuming the initial conditions are known. The model is

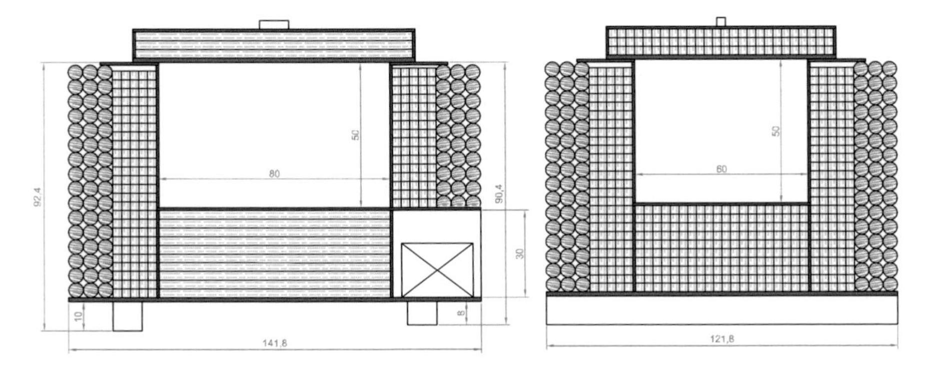

Fig. 3 Technical drawings of the system (upper: longitudinal section, below: transversal section)

based on the energy balances of the internal air and the whole refrigeration system, solved at any desired time step. The thermal inertia of the envelope is considered negligible, since it is realized with lightweight insulation materials.

By analyzing the obtained results from simulations and experimental activities, it was possible to state that:

- the average air temperature measured inside the refrigerator is always below 10 °C, even if a new thermal load is introduced in the course of the completely cloudy days, when the compressor is off. This result confirms the adequate design of the system;
- The error between the temperature calculated and that measured is, on average, lower than $\pm 10\%$ and can be mainly attributed to the impossibility to estimate precisely the heat transfer coefficient of the thermal load inside the refrigerator; this confirms that the developed energy model well describes the behavior of the system and can be used to assess its performances in different contexts.

The trial phase in laboratories lasted 6 months after which the validation phase was carried out within a real context. In detail, the Bangang village located in the northwestern section of Cameroon was selected for installation: there, in 2016, a 4-module (1000 L) solar refrigerator was built (Fig. 4).

As was already mentioned, local materials and man power have been used and involved in order to promote and support the internal economy. In fact, the population has broadly accepted the project and was enthusiastic about participating actively in its construction. This way, an overall cost of the materials and components needed to manufacture one module of the refrigerator (250 L) is assumed to be around 500 € (excluding VAT, shipment costs, and labor).

Such results can ensure a large penetration of the technology since it can be considered immediately affordable for food retailers and small communities.

The project is now facing the data collection phase to understand strengths and weaknesses of the proposal both under a technical point of view and under a social one, while also looking to see if the first signs of feedback from the field application are promising.

Fig. 4 View of the 4-module (1000 L) SPARK prototype built in Cameroon in 2016

4 Conclusions

In the present work, a solar refrigerator kit has been presented by describing its purpose, its technical components, and the benefits provided by its adoption in developing countries. The validation phase carried out both in laboratory conditions and on the field (Cameroon) confirmed a promising effectiveness and efficiency, and also if further improvements are possible. In addition, once again from an economic standpoint, the obtained manufacturing cost is promising, since the high use of local materials and man power helps to keep the refrigerator affordable compared to other similar market products.

Once the proposal has been fully validated and implemented, it would be possible to further test it in other geographical areas and social contexts in order to elaborate a scalability plan aimed at understanding the real replicability of SPARK and where it could be particularly effective. In fact, SPARK has a great potential to enhance food safety and, in particular, to reduce waste along the food supply chain of fish, meat, and several agricultural goods. Moreover, it could also be tested in healthcare facilities, e.g. for preserving vaccines. For now, before spreading its use across the medical field, it would be necessary and advisable to better stress the possible risks that arise from its adoption, which can be assessed in further stages of the research.

Acknowledgements The project was developed in collaboration with the African Center for Renewable Energies and Sustainable Technologies (ACREST), a non-governmental organization located in Cameroon.
The project was funded within the Polisocial Award competition.

References

Capolongo S, Battistella A, Buffoli M, Oppio A (2011) Healthy design for sustainable communities. Annali di igiene: medicina preventiva e di comunità 23(1):43–53
Capolongo S, Buffoli M, Riva MG, Tognolo C, Oppio A (2012) Hygiene and emergency: considerations and proposals for improving hygiene and health features of advanced medical post. Ann Ig 24(5):389–396
Critoph RE, Thompson K (2002) Solar energy for cooling and refrigeration. University of Warwick, Coventry, CV, UK
Del Pero C, Butera FM, Buffoli M, Piegari L, Capolongo L, Fattore M (2015) Feasibility study of a solar photovoltaic adaptable refrigeration kit for remote areas in developing countries. In: 2015 international conference on clean electrical power (ICCEP). IEEE, pp 701–708
Del Pero C, Butera F, Piegari L, Faifer M, Buffoli M, Monzani P (2016) Characterization and monitoring of a self-constructible photovoltaic-based refrigerator. Energies 9(9):749
Del Pero C, Aste N, Paksoy H, Haghighat F, Grillo S, Leonforte F (2018) Energy storage key performance indicators for building application. Sustain Cities Soc 40:54–65
Ewert MK, Bergeron III DJ (2000) Development of a battery-free solar refrigerator. Available online: https://ntrs.nasa.gov/search.jsp?R=20100042308. Accessed on 9 Sept 2016
Ewert MK, Agrella M, DeMonbrun D, Frahm J, Bergeron DJ, Berchowitz D (1998) Experimental evaluation of a solar PV refrigerator with thermoelectric, stirling and vapor compression heat pumps. In: Proceedings of solar, vol 98
Feldman D, Barbose G, Margolis R, Wiser R, Darghouth N, Goodrich A (2012) Photovoltaic (PV) pricing trends: historical, recent, and nearterm projections (No. LBNL-6019E). Lawrence Berkeley National Lab. (LBNL), Berkeley, CA (United States)
Freni A, Maggio G, Vasta S, Santori G, Polonara F, Restuccia G (2008) Optimization of a solar-powered adsorptive ice-maker by a mathematical method. Sol Energy 82(11):965–976
Haver K, Harmer A, Taylor G, Latimore TK (2013) Evaluation of European Commission integrated approach of food security and nutrition in humanitarian context. European Commission Humanitarian Aid. http://ec.europa.eu/echo/files/evaluation/2013/food_security_and_nutrition.pdf
IMECHE (2014) A tank of cold: Cleantech Leapfrog to a more food secure world. Available online: https://www.imeche.org/docs/default-source/reports/a-tank-of-cold-cleantech-leapfrog-to-a-more-food-secure-world.pdf?sfvrsn=0. Accessed on 9 Sept 2016
Kanade AV, Kulkarni AV, Deshmukh DA (2015) Solar power adsorption ice maker system. Int Res J Eng Technol 2:477–486
Kim DS, Ferreira CI (2008) Solar refrigeration options—a state-of-the-art review. Int J Refrig 31(1):3–15
Metcalf SJ, Tamainot-Telto Z, Critoph RE (2011) Application of a compact sorption generator to solar refrigeration: Case study of Dakar (Senegal). Appl Therm Eng 31(14–15):2197–2204
N'Tsoukpoe KE, Yamegueu D, Bassole J (2014) Solar sorption refrigeration in Africa. Renew Sustain Energy Rev 35:318–335
Nemet GF (2006) Beyond the learning curve: factors influencing cost reductions in photovoltaics. Energy Policy 34(17):3218–3232
Rebecchi A, Gola M, Kulkarni M, Lettieri E, Paoletti I, Capolongo S (2016) Healthcare for all in emerging countries: a preliminary investigation of facilities in Kolkata. India. Annali dell'Istituto superiore di sanita 52(1):88–97
Santori G, Santamaria S, Sapienza A, Brandani S, Freni A (2014) A stand-alone solar adsorption refrigerator for humanitarian aid. Sol Energy 100:172–178
Somerton CW, Aslam N, McPhail K, McPhee R, Rowland B, Tingwall E (2009) Vaccine refrigerator for developing nations. Michigan State University, East Lansing, MI, USA
Tassou SA, Lewis JS, Ge YT, Hadawey A, Chaer I (2010) A review of emerging technologies for food refrigeration applications. Appl Therm Eng 30(4):263–276

World Health Organization (2018) The state of food security and nutrition in the world 2018: building climate resilience for food security and nutrition. Food & Agriculture Org

Yildiz A (2016) Thermoeconomic analysis of diffusion absorption refrigeration systems. Appl Therm Eng 99:23–31

Recycling

Introduction

Cinzia Talamo, Niccolò Aste, Corinna Rossi, Rajendra Singh Adhikari

Building materials play a significant role in sustainable architecture. In all tropical countries, traditional construction materials and methods are still used in buildings: some of their advantages are their plentiful supply, low environmental impact, low costs, as well as good reaction to climate; moreover, they can be handled by local labour, which is familiar with both their production and repair. The use of modern building materials, generally imported, is now developing in towns. Modern materials are generally characterised by a high environmental impact, especially as far as the embodied energy is concerned. On the contrary, eco-friendly materials are characterized by low-embodied energy and low emissions, are durable and convenient for recycling and reuse.

In addition, the impact of waste management and disposal should not be underestimated because it affects material and energy savings, shortage of sanitary landfill capacity, ground and water pollution, sanitary conditions, goods production and consumptions. Urban waste cannot be simply considered as a quantity of matter to be disposed: in developing cities, it must be considered first of all as a sanitary problem. The recovery of material and energy from waste should be promoted as much as possible on the basis of the local peculiarities, features and attitudes and properly handled thanks to the available technologies.

Furthermore, life-style and economic systems can directly and indirectly influence energy needs, as well as waste generation and management, and the ensuing important impacts on the environment. All these considerations could also support waste prevention policies based on the redesign of the production and consumption patterns towards low waste and low carbon settlements.

Waste management is multi-sector in nature and encompasses policy making, strategies thinking, the development of legal, institutional, financial and administrative frameworks, as well as the functional design, implementation, operation and management of waste handling facilities.

This section addresses the issues of waste management and recycling in Africa through two different research projects, representative of two different, possible approaches.

Ski Yurt: Upcycle of Downhill Skis for a Shelter in Cacine—Guinea-Bissau

Graziano Salvalai, Marco Imperadori, Marta M. Sesana and Gianluca Crippa

Abstract Downhill skis are composite materials with very high performances. Very often after few years of use, they are changed due to new fashions or ski techniques. This means a lot of waste which still has the potential for very high structural performances. Thus, the idea to upcycle these materials thanks to a cooperation with the University of Grenoble in the joint activities of their laboratory and Velux Lab. Several tests and dome structures were realized in order to show the potential of these materials. Then a yurt was designed, tested, and pre-built at the Politecnico di Milano before shipping it to Africa, through the equatorial forest in Guinea-Bissau, as a first shelter-base camp in a desolated land where a mission was later founded. The purpose is also to make a structure with very high performances and that is resistant to the aggression of termites (only hard wood is suitable there but it requires deforestation) without using steel since it is too expensive. So the final goal is not to send waste to Africa but to show how waste can also become a very solid structure and a valued asset.

Keywords Shelter architecture · Skis · Upcycling · Recycling

1 Introduction

Skiing is one of the most popular sports in the world. According to recent estimates, about one hundred million people ski regularly or occasionally. Out of the European countries, Germany has the most people skiing by far with roughly 14.6 million participating in the sport, followed by France with approximately 8.5 million and the United Kingdom with 6.3 million (www.statista.com). The main equipment consists

G. Salvalai · M. Imperadori (✉)
Architecture, Built Environment and Construction Engineering—ABC Department, Politecnico di Milano, Milan, Italy
e-mail: marco.imperadori@polimi.it

M. M. Sesana
Polo Territoriale di Lecco, Lecco, Italy

G. Crippa
Lecco, Italy

N. Aste et al. (eds.), *Innovative Models for Sustainable Development in Emerging African Countries*, Research for Development, https://doi.org/10.1007/978-3-030-33323-2_7

on skis which, after a number of seasons tend to be disposed of because of delamination, splinters, cracks, etc., or simply because they have become obsolescent. The skis are in general constructed from high-tech materials made up by a sequence of several layers: steel plates, plastics, and resins that gives high resistance and good ductility. The material composition of the skis is particularly complex and they cannot be recycled easily by deconstruction: the making-up materials of a typical alpine ski are assembled with a sandwich structure, perfectly bonded together and hard to disassemble. An alpine ski weighs about 1.8 kg, 35% of which is represented by the angular steel blades, 20% by the wood core, 10% by the surface foil, and the remaining part (about 35%) is mostly represented by adhesives, foams, and rubber elements (Wimmer and Ostad-Ahmad-Ghoradi 2007). Regarding its disposal, the only recycled component is represented by the metal layer, while the remains are shredded and burned in incinerators contributing to the CO_2 emission. Considering the cradle to grave approach (manufacturing, distribution, use, and disposal) and the approximately 1.500 tons of skis fallen into disuse every year, ski manufacturing absorbs a substantial amount of energy. The number of most ski areas in Europe can be found in Germany, with a total of 498. There are 349 ski locations in Italy, 325 in France, and 321 in Russia. The statistics shows a large number of skis sold every year: the number of alpine ski units sold in the USA from the 2015/2016 was close to 750,000. A survey carried out in five large ski areas in northern Italy, between March and April 2014, shows the amount of ski disposals (Table 1).

The interviews involved 23 ski shops with the aims to analyze the local ski's market trend, the technical life of the skis, and the amount of the equipment disposed of. From the data collected, about 2100 skis, approximately three tons of high-tech material have been disposed of only during the winter season of 2014. In this scenario, supported also by the environmental impact analysis, reuse at the end of their life span represents an interesting opportunity. Furthermore, considering the high-performance characteristic concerning both the geometrical and the structural resistance, skis can be efficiently used as structural components in emergency shelter design. Several scientific studies concerning emergency post disaster shelters are available in literature dealing with the technological design, the adaptability, and

Table 1 Analysis of ski disposal in Lombardy (Italy)

Place (Province)	Number of ski's shop interview	Skis disposed per year
Livigno (SO)	2	900
Bormio (SO)	1	300
Ponte di Legno (BS)	3	200
Madesimo (SO)	2	200
Valle Seriana (BG)	15	500
Total	–	2100

Source Surveys carried out by the authors between March and April 2014

versatility of different solutions (Alegria Mira et al. 2014; Crawford et al. 2005; Battilana 2001; Meinhold 2014). Only few publications have focused on the reuse of recycled materials for shelter construction (Imperadori et al. 2014; Salvalai et al. 2017). The purpose of this work is to investigate the reuse of high-tech recycled materials for emergency and temporary architecture exploiting the characteristic of high technology (Daudon 2015; Jalesse et al. 2015). A temporary emergency shelter called "Ski Shelter" composed of recycled skis and covered in a lightweight envelope composed of thermal-reflective-multilayer insulation and polyvinyl chloride (PVC) sheet, has been studied, and built in real scale. The joint between skis has been tested, from the structural point of view, with experimental tests. The first tent prototype has been donated to the Missionaries Oblates of Mary Immaculate and is now operating in the Republic of Guinea-Bissau.

2 The Ski Shelter: Concept and Technology

The Ski Shelter project aims to develop a shelter prototype characterized by an easy assembly method, lightweight, and reused materials, with the capacity to maintain acceptable internal thermal comfort conditions for the user in hot and cold temperatures.

The basic design began from the archetype of the Mongolian yurt, represented by a portable, round tent covered with skins or felt and used as a dwelling by several distinct nomadic groups in the steppes of Central Asia (Fig. 1). The ski yurt prototype has been designed as a composition of a regular grid made up of: 24 concentric axes, representing the beams, and 24 pillars, which together divide the circular base. At the center of the room, a pillar composed of reused standard steel elements for building scaffolding supports the openable 80.0 × 80.0 cm skylight. Linear assembly of several skis constitutes the beams and pillars, the tips of each ski have been previously modified, according to the structural design. The skis have been assembled together to create composite beams and pillars, increasing the inertia and stability, locally and globally. The joints between skis were reinforced with wooden spacer elements (Fig. 2).

The designed geometry has a diameter of 6.0 m, which covers an area of approximately 30.0 m^2 (Fig. 3). The internal height is equal to 1.70 m at the lowest point and 3.80 m in the center of the building area. The shelter is modular and the number of units is potentially implementable according to the emergency situation requirements: the coupling of multiple Ski Yurts allows for a variation of the spaces reaching variable housing dimensions. The building envelope has been created from a textile material coupled with different resistive layers: a double PVC layer, both outside and inside, with a thermo-reflective insulation system interposed in between (Fig. 4).

The high of the pitched roof has been increased in order to maximize the "chimney effect" and improve consequently thermal comfort during summer (Fig. 5). The Ski Yurt prototype can be used as a single shelter or can be assembled into a more complex architecture by means of a modular connection space.

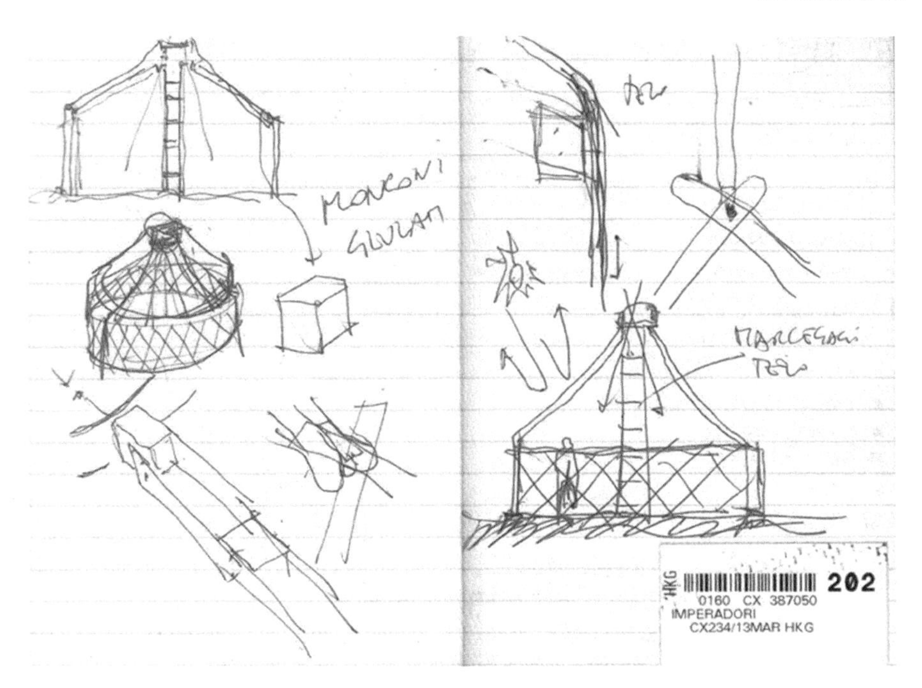

Fig. 1 Schematic concept of the ski yurt. *Source* Marco Imperadori

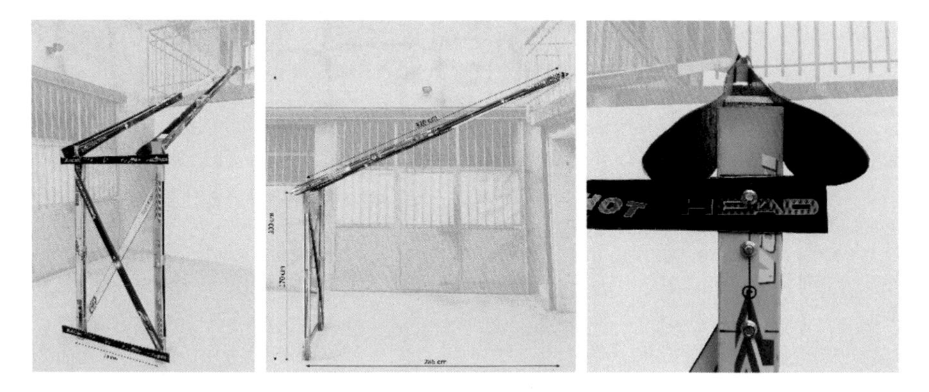

Fig. 2 Photorealistic view of a structure module

2.1 Technological Design

In order to make shelters easy to erect and dismantle, they need to be light weight and have few and easy assembled pieces. Certain types of shelters, such as plastic sheets and tents, are simply erected for a short time span and then dismantled. If the design of a shelter is complex, it will require more training and resources to build it, leading to potential delays (Bashawri et al. 2014). The solutions adopted for the

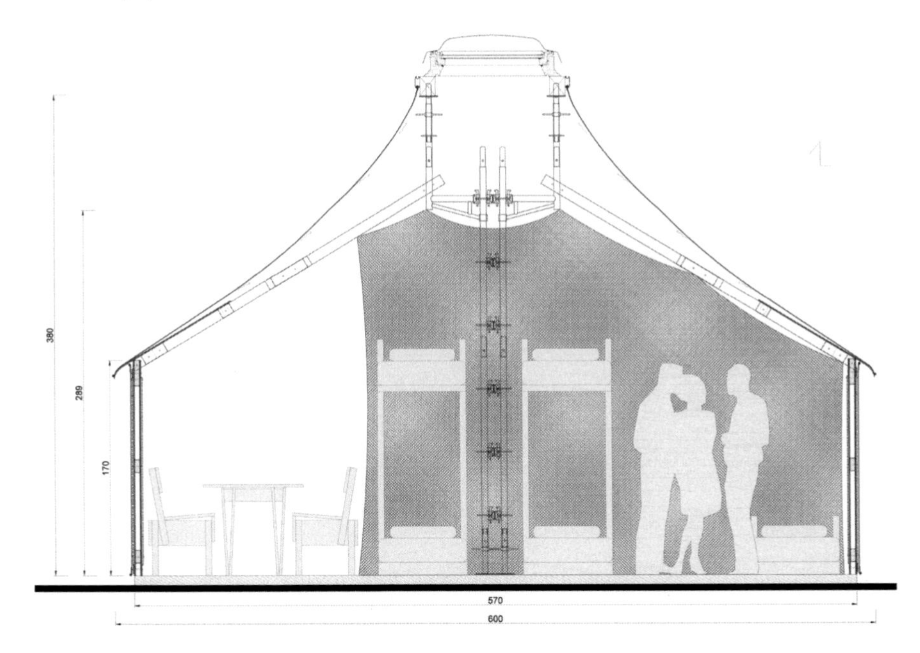

Fig. 3 Vertical section of a standard Ski Shelter

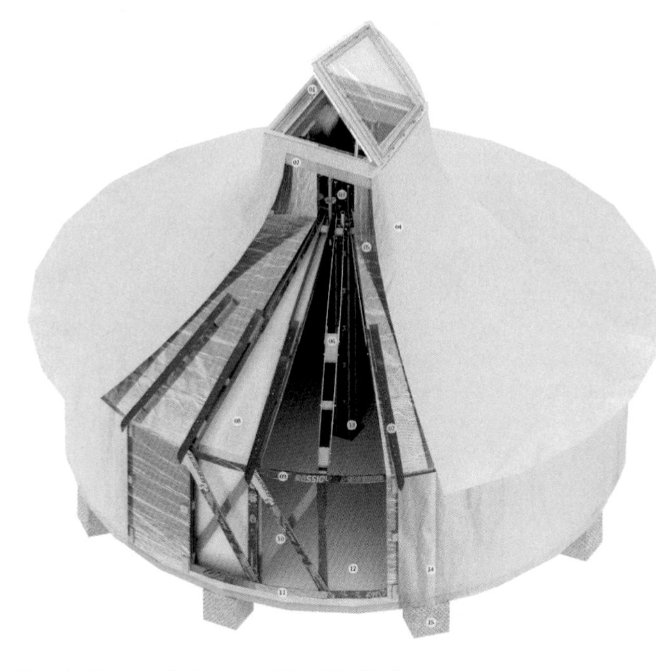

Fig. 4 Photorealistic view of the Ski Shelter

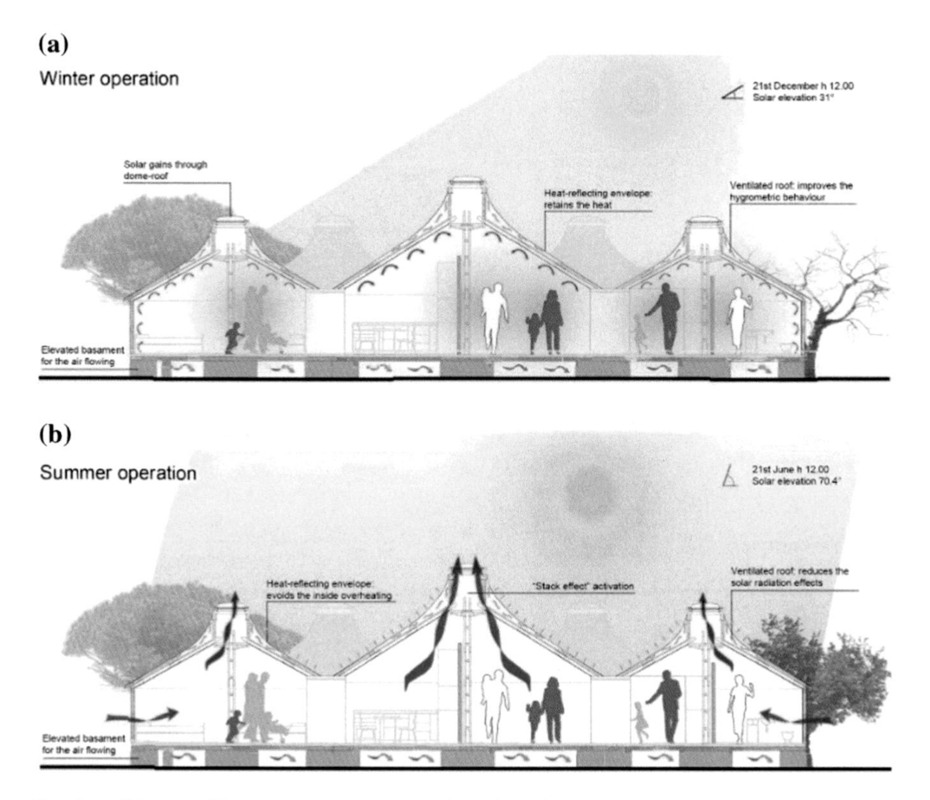

Fig. 5 **a** Winter and **b** summer schematic design of the Ski Shelter

Ski Shelter assembly do not require special skills or in situ operations, and respect the requirements typical of the emergency structure, such as flexibility, portability, lightness, quick installation process, and durability. Considering that the external layer has been made up from a polyester/PVC material, which is characterized by a translucency range of about 0.8–4.0% and with high resistance (Fig. 6). This material presents effective fire resistance and a low specific weight (1450.0 g/m^2). Currently, polyester/PVC is the material most commonly used in architecture since it provides a good balance between cost, performance, and durability (the average life span is between 7 and 15 years), strength and elasticity (Campioli and Zanelli 2009). The thermal resistance to heat transfer is given by the presence of the reflective multilayer insulation (MLI), material already tested by different research studies (Imperadori et al. 2013; Ward and Doran 2005; Salvalai et al. 2015).

This material is a thermal insulation composed of multiple layers of thin sheets developed mainly for spacecrafts. It is commonly used on satellites and other applications in a vacuum where conduction and convection are much less significant, and radiation dominates. In general, the material consists of a series of reflective films covered in a material with low emissivity such as aluminum films and reflective plastic films. For about 30 years, these materials have been used in building, where they

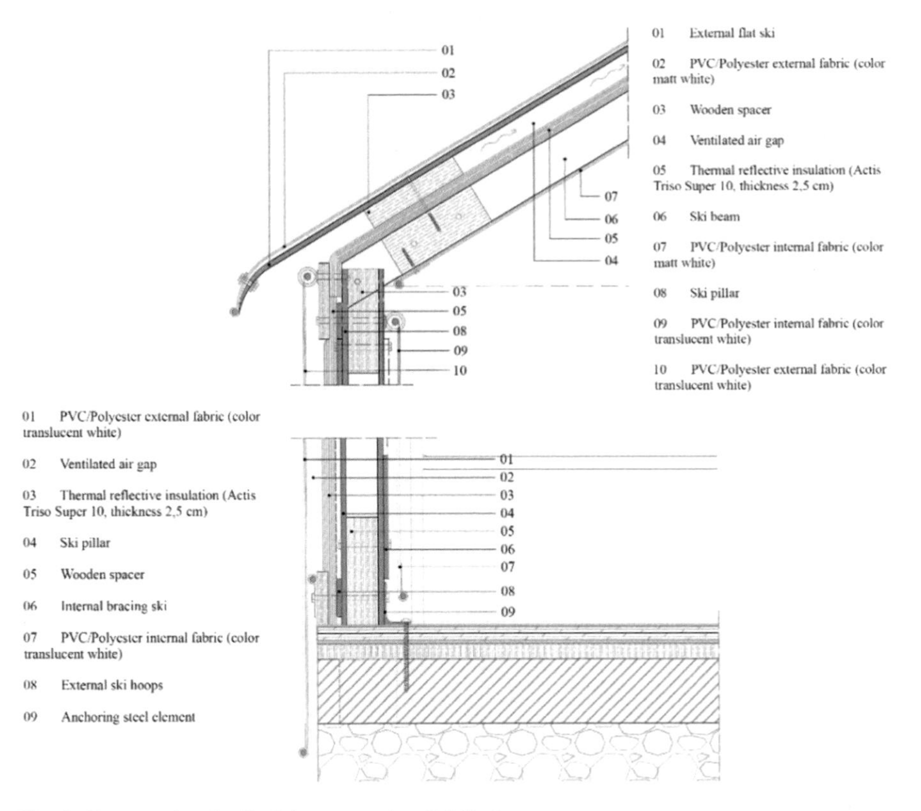

Fig. 6 Constructive detail of the connection Ski Shelter

are able to effectively solve, and in an innovative manner, the problem of thermal insulation, acting in all heat transfer directions. The MLI used in the present study is the ACTIS TRISO Super 10 multilayer (http://www.actis-isolation.com). It is composed of synthetic materials such as wadding sheets or plastic material; and natural fibers, such as sheep's wool. The combination of these thin materials, arranged in succession to one another, gives excellent results in terms of thermal performances: in winter months, it prevents the heat flow from the inside to the outside, while in summer months, when the radiative component of the external thermal load is bigger than the conductive and convective one, it reduces the heat flow from the outside. Figure 6 shows the technological details of the Ski Shelter: the connection between the ground and the skis and between the vertical wall and the roof structure.

2.2 Concept Design and Testing Procedure

The Ski Yurt geometry offers exceptional mechanical resistance being composed of several structural elements close to each other. As mentioned before, the Yurt is built on the basis of 24 concentric triangles with a rotation of 15°. In the center of the Yurt a steel pillar sustains the "crown" which acts as a support for the roof window and the other structural elements that converge on it. To each axis corresponds a composite "ski beam" tilted 30° on the horizontal plane, which is anchored to the "crown" at the top and to the "ski pillar edge" at the bottom (Fig. 7). At both points, the joint does not neither transfer shearing actions nor bending moment (hinge behavior). Different diagonal skis contribute to withstanding the horizontal forces. In the end, the horizontal stress coming from the beams is absorbed by two orders of "ski hoops" placed one on the top and one on at the base of the pillars' edge. Again, these joints are discretized as three-dimensional hinges. All the nodes of the structure have been connected by threaded steel bolts.

In 2015, together with the Politecnico di Milano and ESPE Lecco (a technical professional school for construction workers in Lecco), the Ski Yurt was tested verifying the construction phases. The main goal was to derive specification and suggestion for further design and construction optimization. During the test, all the technical elements have been numerically named to help and speed up the assembly phase (Fig. 8).

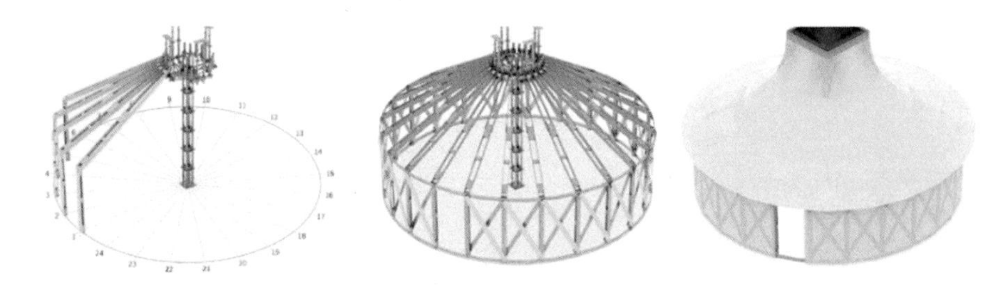

Fig. 7 Technical drawing of the different construction phases

Fig. 8 Pictures of the pre-assembly phase

The whole construction process took four working days to assemble the skis for the individual structural elements, such as beams and pillars. Then, it took an additional five days to assemble the whole yurt structure. The working team was composed of an average of four people, two of them skilled workers and two of them students. The shelter was then taken apart in half a day, packed and loaded into a container ready for shipping to Guinea-Bissau.

3 Performance Verification: Simulation Study

Several studies are available in literature analyzing the shelter performances through experimental and simulation studies in different geographic areas that verify the internal conditions (Cornaro et al. 2015; Ajam 1998; Manfield 1999, 2000). The tent prototype was also analyzed by dynamic thermal analyses performed with Trnsys v.17 Environment (https://sel.me.wisc.edu/trnsys/). The geometrical model (Fig. 9) was modeled in Trnsys3d (http://www.trnsys.de/), a plug-in for Google SketchUp.

The geometry was modeled as a homogenous thermal zone with a total volume of 83.15 m^3. The vertical surface (33.84 m^2) and pitched roof (43.15 m^2) have been implemented in type 56 considering a solar absorbance coefficient of 0.4 (clear color) and the convective heat transfer coefficients were set equal to 3.0 W/(m^2°C) for internal surfaces and 17.8 W/(m^2°C) for the external ones. The ground floor was modeled considering the direct contact between the ground and the internal zone. The model allows for predicting the internal temperature considering in one side the free-running operation and in the other side the ideal heating/cooling demand with an internal set point control of 18 °C in winter and 26 °C in summer. The simulation was performed considering the climate of Palermo (Italy), characterized by high ambient air temperature in the summer with high solar heat gains. The thermal performance of the tent is summarized in the Table 2.

Different ventilation strategies have been tested: night-time ventilation (8:00 p.m. to 8:00 a.m.) and day-time ventilation (10:00 a.m. 14:00 p.m.) both with an air change of 3 volumes per hour (ach). Changing from day ventilation to night ventilation there is no significant reduction of the temperature level due to the low thermal inertia of the yurt envelope. In general, intensive ventilation strategies allow for an increase in the

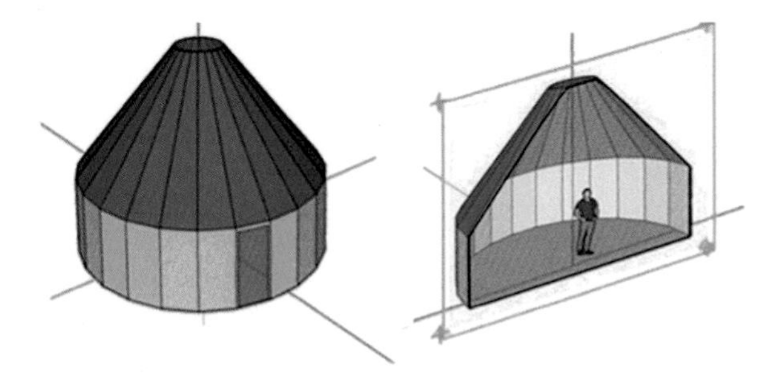

Fig. 9 Three-dimensional single thermal zone simulation model

Table 2 Features of the elements that make up the multilayer technology

	Thickness (mm)	R (m^2K/W)
Wall/roof ($R_{tot} = 5.538$ m^2K/W)		
R_{si}		0.131
Internal envelope in Polyester/Pvc	1.20	0.006
Air gap	60	0.182
Insulation ACTIS TRISO SUPER 10	25	5
External envelope in Polyester/Pvc	30	0.176
R_{se}		0.038
Basement ($R_{tot} = 3.768$ m^2K/W)		
R_{si}		0.171
Internal envelope in Polyester/Pvc	1.20	0.006
Background slab	25	0.834
Insulation ACTIS TRISO SOLS	7	2.5
OSB panel	28	0.215
R_{se}		0.038

interior comfort conditions. The thermal comfort level has been analyzed according to the EN UNI 15251:2008 standard. Figure 10 shows the correlation between the operative room temperature (ORT) and the running mean ambient air temperature.

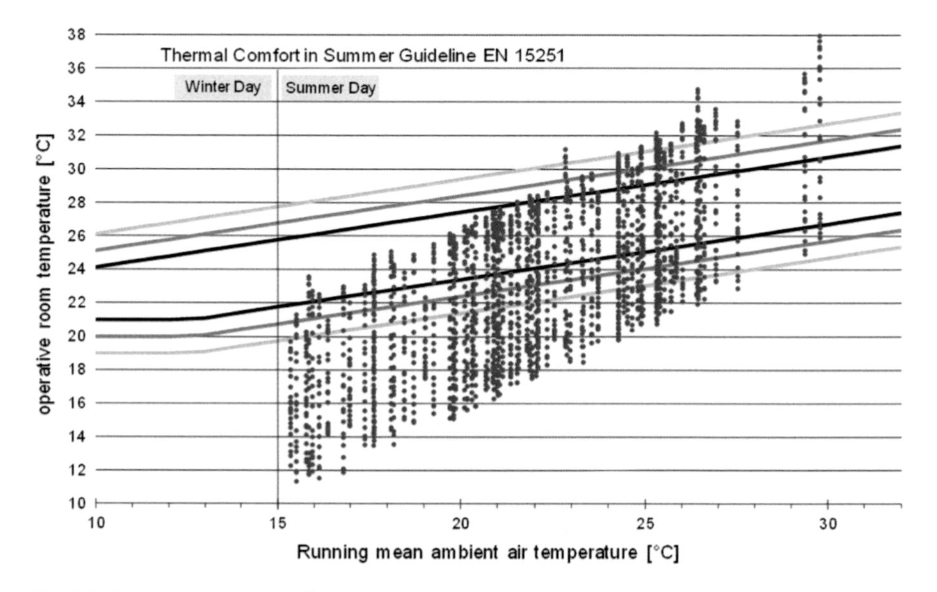

Fig. 10 Summer thermal comfort evaluation according to the EN 15251 standard

Fig. 11 Construction phases in Cacine, Guinea-Bissau

The annual energy balance (useful energy) shows that the cooling consumption exceeds that for heating at about twice the amount: the energy for cooling is equal to 1003 kWh and those for heating equal to 559 kWh. The total energy related to the floor area is equal, respectively, to 33.3 kWh/m^{2}y and 18.6 kWh/m^{2}y, reflecting the predominant summer climate conditions of southern Italy. The low energy-consumption level would lead to different benefits such as energy procurement and fuel transportation.

4 Conclusion and Development

The presented work demonstrates the high potential of upcycle materials like disposed of skis, in order to use them for building temporary and emergency architectures, due to their features of lightness, quick installation, and flexibility.

The Ski Yurt shows an innovative concept of potential living, a secure, comfortable, and healthy space easily assembled and disassembled. Modularity and simple connections between different components guarantee easy transport, assembly, and maintenance by locals without specific tools and skills. The experimental Ski Yurt, after the assembly test in Italy, has been donated and transported to the Cacine community to become the base camp for the new mission center coordinated by the Oblate Fathers of Mary Immaculate (Fig. 11). Local people rebuilt the Ski Yurt easily, and since then it has become an important reference point for the population as a useful, comfortable, and safe space.

Acknowledgements The presented work has been carried out on the basis of the content of the master thesis entitled: ARCA project. Architecture of resilience and community accommodation. Android—resilience shelter: strutture temporanee per l'emergenza abitativa, resilienti e ad elevate prestazioni termodinamiche (authors: Federico Lumina, Elisa Mutti, Ilaria Polese). The authors would like to thank the students for the amazing work and for the effort in making possible the construction of the Ski Yurt prototype. The authors would like to thank all the industrial partners who provided materials and expertise: Velux, Cittadini, SergeFerrarì, Canobbio, Marcegaglia, Joseph Fourier University.

References

Ajam R (1998) Thermal comfort in low-cost refugee shelters: a computer simulation study in Waqas, a lower desert valley area in Jordan. UNRWA-HQ, Amman
Alegria Mira L, Thrall AP, De Temmerman N (2014) Deployable scissor arch for transitional shelters. Autom Constr 43:123–131
Bashawri A, Garrity S, Moodley K (2014) An overview of the design of disaster relief shelters. Proc Econ Finance 18:924–931
Battilana R (2001) Design of cold climate temporary shelter for refugees. University of Cambridge. Master thesis. Source: www.shelterproject.org/downloads/cold%20climate%20liner.pdf
Campioli A, Zanelli A (2009) Architettura tessile. Progettare e costruire membrane e scocche, Il Sole 24 Ore, Milano
Cornaro C, Sapori D, Bucci F, Pierro M, Giammanco C (2015) Thermal performance analysis of an emergency shelter using dynamic building simulation. Energy Build 88:122–134
Crawford C, Manfield P, McRobie A (2005) Assessing the thermal performance of an emergency shelter system. Energy Build 37:471–483
Daudon D (2015) Innovation and sustainable development in civil engineering degrees: constructing structures with re-used skis. Université Joseph Fourier, Grenoble
Imperadori M, Pusceddu C, Salvalai G (2013) Thermal-reflective multilayer insulation systems in the emergency architecture: the Air Shelter Skin, Plea Conference 2013. Munich, Germany
Imperadori M, Salvalai G, Pusceddu C (2014) Air shelter house technology and its application to shelter units: the case of scaffold house and cardboard shelter installations. Procedia Econ Finan 18:552–559
Jalesse K, Wauthy R, Diagne A (2015) Innovation et Développement Durable dans les formations GC: conception collaborative et normative pour l'habitat d'urgence, réalisation d'une yurte en skis usages. Joseph Fourier University, Grenoble
Manfield P (1999) A comparative study of temporary shelters used in cold climates. Martin Centre
Manfield P (2000) Modelling of a cold climate emergency shelter: prototype and comparison with the United Nations Winter Tent. Martin Centre, Cambridge
Meinhold B (2014) Urgent architecture: 40 sustainable housing solutions for a changing world. W. W. Norton & Company, Inc
Salvalai G, Imperadori M, Scaccabarozzi D, Pusceddu C (2015) Thermal performance measurement and application of a multilayer insulator for emergency architecture. Appl Therm Eng 82:110–119
Salvalai G, Imperadori M, Lumina F, Mutti E, Polese I (2017) Architecture for refugees, resilience shelter project: a case study using recycled skis. Procedia Eng 180:1110–1120
Skiing in Europe—Statistics & Facts, https://www.statista.com/topics/3922/skiing-in-europe/
Thermal reflective insulation materials. Source: http://www.actis-isolation.com
UNI EN 15251:2008 Criteri per la progettazione dell'ambiente interno e per la valutazione della prestazione energetica degli edifici
Ward TI, Doran SM (2005) Thermal performance of multi-foil insulation, Report BRE, Scotland, Glasgow. Source: www.planningportal.gov.uk/uploads/br/multi-foil-insulation_july2005.pdf
Wimmer W, Ostad-Ahmad-Ghoradi H (2007) Ecodesign of Alpine skis and other sport equipment—considering environmental issues in product design and development. The Impact of Technology on Sport II, Taylor & Francis Group, pp 15–23

Comprehensive Feasibility Study for the Construction of an Integrated Sustainable Waste Management Facility in Kajiado County, Kenya

Claudio Del Pero, Niccolò Aste, Paola Caputo, Francesca Villa and Mario Grosso

Abstract The work focuses on the feasibility study relating to the design of a new integrated sustainable waste management facility in Ngong town (Kenya), within the Kajiado County; the County borders Nairobi and extends to the Tanzanian border further south. Currently, the waste generated in Ngong town is sent to an illegal dumping site, which must be closed as soon as possible since it is causing serious environmental and social impacts. Specifically, the purpose of the study was to carefully analyze the context, taking into consideration environmental, economic and social aspects, providing technical–economic solutions which must be robust, easy to build, operate and maintain, be cost-effective and self-sustaining from an economic standpoint, and be environmentally and socially sustainable. This chapter focuses on the assessment carried out in view of designing a new integrated sustainable waste management facility. An estimation of the mass and energy balance and of the system design and sizing is also provided, together with an assessment of the overall economic sustainability of the proposed intervention.

Keywords Integrated waste management · Renewable energies · Developing countries

1 Introduction to Proper Waste Treatment Strategies

The present chapter focuses on the feasibility study relating to the design of a new integrated sustainable waste management facility in Ngong town (Kenya), within the Kajiado County; the County borders Nairobi and extends to the Tanzanian border further south. Currently, the waste generated in Ngong town is sent to an illegal dumping site, which must be closed as soon as possible since it is causing serious environmental and social impacts.

C. Del Pero (✉) · N. Aste · P. Caputo
Architecture, Built Environment and Construction Engineering—ABC Department, Politecnico di Milano, Milan, Italy
e-mail: claudio.delpero@polimi.it

F. Villa · M. Grosso
Department of Civil and Environmental Engineering—DICA, Politecnico di Milano, Milan, Italy

N. Aste et al. (eds.), *Innovative Models for Sustainable Development in Emerging African Countries*, Research for Development, https://doi.org/10.1007/978-3-030-33323-2_8

The first stage of the project was a decision-making process carried out to identify the most suitable solution for the construction of a waste management facility in Ngong town (Kenya) and subsequently to provide the main design details. First, different modern waste treatment strategies were analyzed, in order to determine the best option, taking into account the peculiarities of application in developing countries (UNEP 2013; UN-HABITAT 2010; USAID 2009; UNEP II 2013; Mohammed et al. 2013). The results of the assessment are summarized in Table 1.

Table 1 Pros and cons of the main usable waste treatment strategies

PROS	CONS
(1) *Thermo–chemical technologies*	
• Robust technology • Maximum reduction in waste mass and volume • Recovery of electricity and/or thermal energy • Waste is sanitized and sterilized • Possibility of recovering materials such as metals during the process	• Very expensive (construction) • Requires high technical skills for operating • Expensive operation because it requires chemicals for flue gas cleaning • Some solid residues generated are hazardous waste (i.e. filter ash) • Heavy maintenance required to guarantee adequate environmental standards • Requires waste with high low-heating value (i.e. with low organic fraction content) as input • In developing countries, it requires auxiliary fuel (i.e. coal) to support combustion • Pyrolysis and gasification are possible only if these thermal treatments are coupled with integrated waste management systems of small regions • Does not allow material recovery by manual operators in proper hygienic conditions
Preferred application Big cities in developed countries with relatively cold climate, where combined heat and power (CHP) can be implemented and heat can be used for district heating purposes	
(2) *Sanitary landfill*	
• Relatively cheap solution, if post-closure costs are not considered • Some energy recovery can take place by exploiting the landfill gas	• Requires large spaces • Does not allow material recovery by manual operators in proper hygienic conditions • Potential release of methane in the atmosphere (greenhouse gas) • Expensive if post-closure costs are considered (at least 30 years of maintenance and monitoring after the useful life) • Risk of open dump sites around the sanitary landfill
Preferred application Sanitary landfill is not an advisable solution, but it can be an option for very low-income countries as a basic alternative to open dumps	

(continued)

Table 1 (continued)

PROS	CONS
(3) *Bio-drying with landfill bioreactor*	
• Reasonable construction costs • Important reduction in water content in the waste, significantly reducing the amount of waste to be landfilled • It allows material recovery by manual operators in better hygienic conditions compared to open dumps and incineration • Very simple operation, not requiring high technical skills • Reasonable maintenance costs and efforts • It allows energy recovery in the landfill bioreactor • It can be converted into a composting plant in case source separation of food waste is established • Very suitable for waste with high water and organic content	• It requires careful management of the landfill bioreactor, to avoid risks of fires and explosions
Preferred application Very flexible technology, which can be applied in a wide range of contexts	
(4) *Anaerobic digestion*	
• Anaerobic digestion allows the production of fertilizers to be used in agriculture • Anaerobic digestion produces biogas to be used in several ways and generate income	• An adequate separation stage for the organic fraction is needed. • Material recovery by manual operators in proper hygienic conditions is not possible. • Potential release of methane in the atmosphere (greenhouse gas)
Preferred application When a high amount of controllable organic waste is available and the digestate has a good local market	
(5) *Composting*	
• Simple and inexpensive technology • Composting allows for the production of high-quality fertilizer (compost) • Small-, medium- and large-scale schemes are available	• Accurate source separation for the organic fraction is needed • The downstream market for compost is not always available, especially if the compost is of low quality • Intense odors might be generated if the exhaust air is not properly treated
Preferred application When a high amount of controllable organic waste is available and the compost has a good downstream local market	

2 Identification of the Most Suitable Solution for the Waste Management Facility

The choice of the most appropriate technology for waste treatment depends on several factors; of these, waste composition is the first aspect to be considered. Within the present collection system in Ngong, mixed waste is collected without any source separation. Consequently, the waste delivered to the facility has a very high moisture content (about 70%), directly connected to the prevalence of organic matter in the waste.

As a consequence, the application of all thermo–chemical technologies is clearly precluded. Moreover, they are expensive and difficult to operate in developing countries, thus were not considered feasible. Regarding the capital cost, it was estimated that an incinerator with the required size for the Ngong project will need an investment of around 50,000,000 €, which is double the cost of the other analyzed technologies. Finally, the technology would hardly allow the re-employment of all the people currently working in the dumpsite.

On the other hand, a sanitary landfill, even if built in compliance with modern standards, has not been considered to be a valid alternative in the long term. In fact, as mentioned previously, this solution will preclude the separation of the recyclable fraction of waste. In this sense, it must be noted that the manual separation of waste as-is was not considered feasible since it cannot be managed in proper hygienic conditions. Food waste being the most relevant fraction (roughly 70% of the overall weight), priority should be given to processes capable of reducing its putrescence and water content. The latter is in the range of 70–80% in weight of food waste.

Having considered all these factors, bio-drying represents the most suitable technology for waste treatment in Kajiado. The suitability of this technology for waste with a high water content has been discussed by several authors throughout scientific literature, as reported, e.g. in He et al. (2013), Tambone et al. (2011), Rada et al. (2007), Tom et al. (2016), Rada et al. (2009), Velis et al. (2009). Mixed waste is processed in a plant that will reduce its putrescence and water content before being delivered to a sorting stage aimed at recovering potentially recyclable materials. An effort toward separate collection would be much more challenging in such a context (also requiring a proper distribution of information and training to the public), but can deliver long-term benefits, such as allowing the recovery of food waste (for animal feed production and for use in agriculture following a composting process) and a better recovery of recyclable materials.

The remaining waste after bio-drying and sorting can be finally disposed of in an engineered landfill bioreactor[1] for biogas production or used as a refuse-derived fuel (RDF) in industrial thermal processes (e.g. cement production), depending on its chemical and physical properties. For the time being, the first option has been taken into consideration, and the conceptual framework of the proposed waste-to-energy strategy is represented in Fig. 1.

[1]https://www.epa.gov/landfills/bioreactor-landfills.

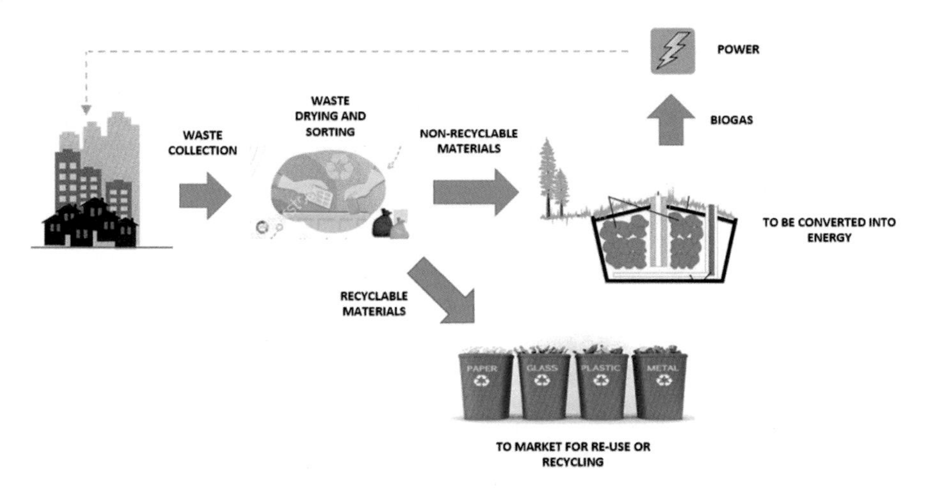

Fig. 1 Diagram of the proposed waste-to-energy strategy

Non-technological issues also need to be taken into account in the final decision-making stages. Specifically, economic sustainability is a key issue. Daily operational and management costs of the facility should be covered by the budget of the Department of Environment and Natural Resources of the Kajiado County. These costs will be in addition to the existing costs relating to waste collection and public hygiene. Bio-drying is among the most affordable and flexible technological options (in comparison with anaerobic digestion or incineration), and it has been proposed in combination with biogas production so that partial self-sustainability may be achieved.

Another aspect that needs to be considered is the organization of the waste management system. Bio-drying is appropriate for the current system, and it can also be adapted for a gradual shift toward separate collection (for example the bio-drying reactor can be used for composting the source-separated food waste). Nonetheless, other management options would require different technologies, along with significant involvement on behalf of public bodies. This is a governance issue which should also be discussed with local institutions and other pertinent stakeholders.

In the short- and medium-term, a bio-drying plant followed by manual sorting and by a landfill bioreactor has been chosen since it presents the following key strong points:

- it allows for the manual separation of mixed waste under more hygienic conditions, enabling the employment of people from the informal sector currently working on the dumpsite;
- it minimizes odor emissions and their relevant social impacts;
- it is a flexible solution which will also be suitable for the future source-separated waste, when a proper separate collection system will have been implemented;
- the capital cost is affordable, especially compared to alternatives such as waste incineration;
- it is a robust technology, without the need for complex high-tech equipment and skills.

3 Description of the New Facility

As already highlighted, bio-drying is particularly suited for processing waste with a high moisture content, since it allows for the partial evaporation of water by using the heat released by the aerobic biological degradation of the organic matter. The only technological interventions required for the process are a preliminary light shredding (aimed at opening the bags in which waste is contained) and forced air intake achieved through the use of simple air blowers, for a duration that can range between 10 and 20 days. This process can be simply operated by non-specialist staff and must be as automated as possible. The staff will only be required to oversee the activity based on a simple and intuitive interface and a few controls. Bio-drying results in a significant reduction in weight and moisture of the waste entering the process (in the range of 25–35%). The exhaust air needs to be treated before being released into the atmosphere. This can be done by using a bio-filter installed on the roof or on the side of the bio-drying building.

In the current case-study, since a solar photovoltaic (PV) plant will be installed on the roof, the bio-filter will be placed on the ground close to the bio-drying building. The bio-filter allows air to pass through retaining and bio-degrading pollutants. The bio-filter will be composed of organic filtering material, such as wood chips or coconut shell fragments, depending on the local availability of suitable materials. The filtering material needs to be periodically integrated and replaced: its life span depends on several parameters that must be considered in the final detailed design of the system, depending on the selected commercial technology (Fig. 2).

Fig. 2 View of a bio-drying system located in Italy, similar to the one described

By reducing the putrescence and humidity of waste, the following sorting operations aimed at recovering potentially recyclable materials (plastic, glass, metals) are simplified and can be carried out with simple mechanical devices (sieves, magnets) or even by hand sorting, thus ensuring the involvement of the existing informal sector. The last option appears particularly suitable for the re-integration of the people currently employed at the dumpsite.

After sorting, residual waste can finally be disposed of in the engineered bioreactor for biogas production or used as a refuse-derived fuel (RDF) in industrial thermal processes (e.g. cement production), depending on its chemical and physical properties.

The bioreactor will take the place of the conventional engineered landfill, receiving residues from previous stages. This must be accurately designed in order to avoid groundwater and soil contamination and requires a drainage system for the leachate (the liquid part leaching from the waste) as well as a collection system for the biogas produced by anaerobic processes. State-of-the-art technologies for achieving such purposes are those in compliance with current European Union legislation on landfilling (Directive 1999/31). Since leachate and biogas production are mainly influenced by the presence of the organic fraction, bio-drying can affect both of these aspects positively by achieving a partial degradation of the organic fraction.

By considering an estimated daily production of 130 tons of waste in the area under study, and an annual 6% increase in waste generation (mainly due to the projected population increase), the new bioreactor will need to accommodate an annual range of 30,000–62,000 tons of bio-dried material (estimated quantities, respectively, for the first and twentieth year). Assuming a typical level of compaction (0.8 t/m^3) and adding the daily coverage material, the total volume in a 20 years' time-span will be about 1,200,000 m^3. This volume may change (hopefully by decreasing) if and when new waste management strategies are put in place, namely the introduction of source separation of food waste. The surface area occupied by the bioreactor will be around 60,000 m^2.

Generation of biogas will start after 1–2 years, since a minimum amount of waste is needed before enhancing its generation by means of leachate recirculation. The amount of biogas produced will be burned in a co-generator unit producing heat and power. The electric energy will be enough to power the bio-drying facility, and the excess generation can be fed to the grid, helping to ensure the economic self-sustainability of the system as a whole.

Despite the fact that managing the bioreactor is aimed at maximizing the recirculation of leachate in order to enhance biogas generation, a certain amount of leachate will be produced and will need to be disposed of. It is very difficult to estimate the exact amount of leachate produced, since it depends on many factors, particularly on the local climatic conditions as well as on the actual management of the bioreactor itself (proper daily coverage, extent of recirculation, etc.). Although some experiences demonstrate the possibility of achieving a very low production of leachate, the literature data is more variable, ranging from 0.02 to 0.26 m^3 per ton of waste disposal. In our estimate, we have conservatively assumed an initial generation of

$0.1\ m^3/t$, targeting a progressive decrease thanks to continuously improved management of the bioreactor.

Following the above-described preliminary sizing of the system, an indicative layout of the plant was developed, as shown in Fig. 3.

Because of the relatively low amount of leachate, and of its possible reduction, it is not advisable to build a dedicated treatment plant at the site considering the complex management of such systems and the lack of rivers or water basins in which to convey the treated water. The excess leachate will be periodically evacuated by means of tank trucks and properly disposed of at wastewater treatment plants in the

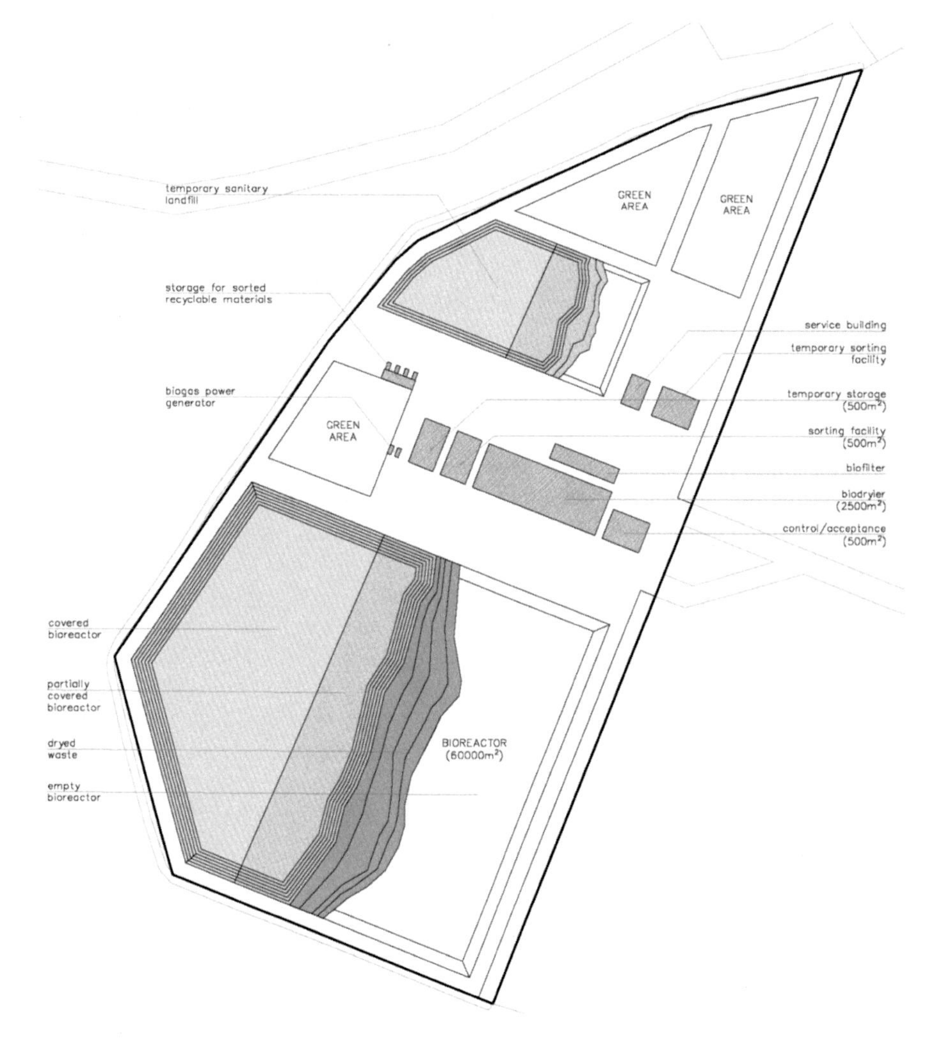

Fig. 3 Indicative system layout (elaboration by the authors)

Nairobi area. A feasibility analysis was carried out and some private companies that are potentially suitable for the purpose were identified.

In addition, in order to ensure economic sustainability, the exploitation of renewable energy sources, specifically PV technology, is recommended in the new facility; This will allow for the generation of electricity to cover the energy demand before the biogas production starts and to minimize the running costs during the entire system's lifetime. All the roofs of the facility's buildings will be covered in a PV system generating a total estimated power around 800 kW_p. It was assumed to have around 5800 m^2 of roof surface and that 1 kW_p of PV modules needs around 7 m^2 of useful surface. Such a PV plant is expected to generate approximately 1,300,000 kWh/year of electricity. Thus, this project also represents an interesting case of effective matching between two renewable and local energy sources.

The biogas collected from the bioreactor will be burned in specific biogas CHP units generating both thermal energy and electricity. Stable biogas production will start 18–24 months after the first operation of the plant. Since the production of electricity from the combustion of biogas is expected to be about 180 kWh per ton of waste entering the bioreactor, a potential production of about 5000 MWh is expected from the third year of plant operation. In 20 years, considering the increase of the daily waste production, it will be possible to reach a potential production of about 10,000 MWh per year.

A flow chart detailing the rounded mass balance of the system in the third year of its operation is shown in Fig. 4.

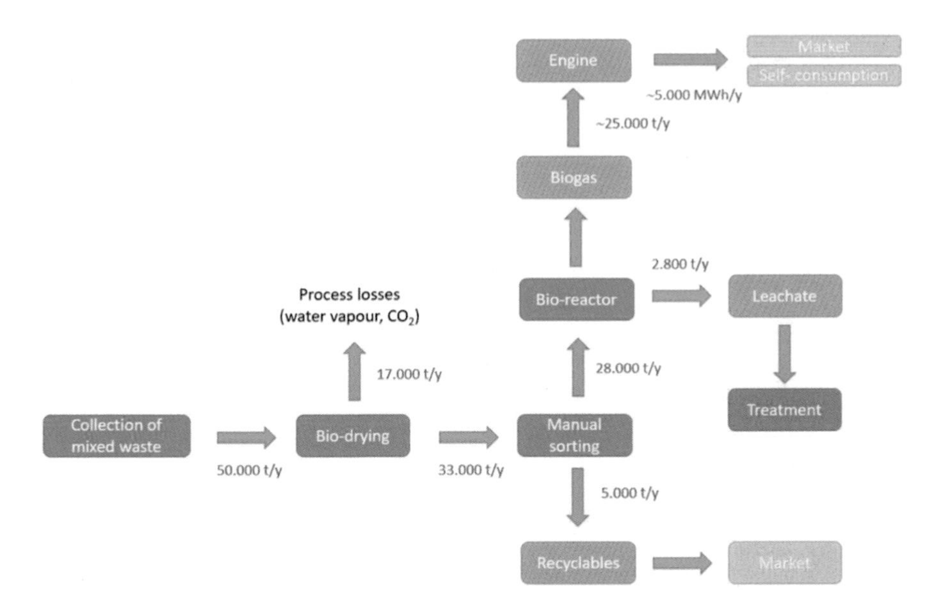

Fig. 4 Overall mass balance of the system at year 3 (rounded figures)

4 Final Evaluations and Conclusions

By considering the information collected during the feasibility study, it was possible to state that the implementation of the new integrated sustainable waste management facility is feasible with the use of robust technologies composed of a bio-drying unit coupled with a bioreactor for the production of biogas. This was assumed as the best technological option for the Kajiado County's context.

The total budget required for the entire intervention (construction of the integrated waste management facility, including the closure of the existing dumpsite and capacity building/training activities) is approximately equal to 20,000,000 €. It must be noted that such an amount must be considered as an average educated guess, pending the huge uncertainties associated with these types of activities, where unexpected events can arise at any time. The new facility (waste-to-energy system), together with the rooftop solar energy system, will be able to generate more than 5,000,000 kWh/year at full power (at around the third year of operation), which is equivalent to an economic counter value ranging from 350,000 to 450,000 €/year. Part of this energy will be used for the operation of the facility, while the excess will be sold to the national grid to generate income. In this way, before the tenth year of operation, the system will begin to generate positive incomes compared to the running costs. The project herein described represents an interesting case of technological transfer from Europe to Africa, taking into account the local peculiarities that strongly affect waste management. Several socio-economic issues have been faced, especially in relation to local population and environment. By monitoring the developments of this experience, future studies will potentially be able to contribute also in terms of technical replicability.

References

He P, Zhao L, Zheng W, Wu D, Shao L (2013) Energy balance of a biodrying process for organic wastes of high moisture content: a review. Dry Technol 31:132–145

Mohammed YS, Mustafa MWn, Bashir N, Mokhtar AS (2013) Renewable energy resources for distributed power generation in Nigeria: a review of the potential. Renew Sustain Energy Rev 22:257–268

Rada E, Franzinelli A, Taiss M, Ragazzi M, Panaitescu V, Apostol T (2007) Lower heating value dynamics during municipal solid waste bio-drying. Environ Technol 28:463–469

Rada E, Ragazzi M, Panaitescu V (2009) MSW bio-drying: an alternative way for energy recovery optimization and landfilling minimization. UPB Sci Bull Ser D 71:113–120

Tambone F, Scaglia B, Scotti S, Adani F (2011) Effects of biodrying process on municipal solid waste properties. Bioresour Technol 102:7443–7450

Tom AP et al (2016) Biodrying process: a sustainable technology for treatment of municipal solid waste with high moisture content. Waste Manage 49

UN-HABITAT (2010) Collection of municipal solid waste in developing countries. United Nations Human Settlements Programme (UN-HABI-TAT), Nairobi, Kenya

UNEP (2013) Municipal solid waste: is it garbage or gold? Available at: UNEP Global Environmental Alert Series. http://www.unep.org/pdf/UNEP_GEAS_oct_2013.pdf

UNEP II (2013) Municipal solid waste management. Available at: Newsletter and Technical Publications. http://www.unep.or.jp/Ietc/ESTdir/Pub/MSW/index.asp
USAID (2009) Environmental guidelines for small-scale activities in Africa: environmentally sound design for planning and implementing development activities. Chapter 15 Solid waste: generation, handling, treatment and disposal. U.S. Agency for International Development. http://www.encapafrica.org/egssaa.htm
Velis CA et al (2009) Biodrying for mechanical–biological treatment of wastes: a review of process science and engineering. Bioresour Technol 100

Environment and Emergency

Introduction

Cinzia Talamo, Niccolò Aste, Corinna Rossi, Rajendra Singh Adhikari

Environmental emergencies can cause severe environmental damage as well as loss of human lives and property. They may result from natural, technological or human-induced factors, or from their combination. Being ready to face these events is essential in order to reduce impacts on human health and environment, and to allow a more effective response and recovery.

The construction industry and the professionals involved in the management of the built environment play a significant role in both pre-disaster risk reduction and post-disaster handling. Every year, natural disasters cause a substantial amount of damage throughout the world. The evaluation of the risk and the operations aiming at mitigating it play a significant and increasingly important role: the construction industry has a strong relationship with disaster management and can, therefore, contribute in a significant way towards the reduction of risks. Risk reduction strategies can be incorporated at all phases of construction projects, including planning, designing, construction and maintenance in addition to town and country planning and policy making.

This section addresses the issues of environmental emergency in the built environment and illustrates examples of feasible solutions that have been envisaged by select research project.

Borboleta and Papagaio: Emergency Unit and Children's Nutritional Center in Farim—Guinea-Bissau

Marco Imperadori, Graziano Salvalai, Marta M. Sesana, Serena Rosa and Consuelo Montanelli

Abstract Farim is a city on a deep fjord of the Atlantic Ocean in Guinea-Bissau. The Mission of Padri Oblati di Maria has operated there for many years and the research team has realized two units in recent years: Borboleta, which is an emergency unit for children but also for all those in need, and Papagaio, which is a nutritional center for young children. Due to the salt of the fjord (which also creates an economy), the atmosphere of the environment is typically saline so steel structures have been protected with special nanotechnologies (thanks to Triplex tech by NordZinc) for Borboleta. Papagaio is a nutritional center which provides food for the youngest groups of children who otherwise would have very few possibilities of survival. The structure of the pitched roof space has been developed with a transfer of technology from industrial scaffold elements transformed into columns and beams. Only one section becomes the rule of construction and the whole structure and sandwich panels roof were built in two weeks by volunteers connected to the Padri Oblati di Maria. After the mechanical erection, the envelope (realized in crude earth blocks and plaster) was completed by local people who in those years had been trained to learn basic masonry rules.

Keywords Steel construction · Recycling · Local materials · Daylighting

M. Imperadori (✉) · G. Salvalai
Architecture, Built Environment and Construction Engineering—ABC Department, Politecnico di Milano, Milan, Italy
e-mail: marco.imperadori@polimi.it

M. M. Sesana
Polo Territoriale di Lecco, Lecco, Italy

S. Rosa · C. Montanelli
Lecco, Italy

N. Aste et al. (eds.), *Innovative Models for Sustainable Development in Emerging African Countries*, Research for Development, https://doi.org/10.1007/978-3-030-33323-2_9

1 Guinea-Bissau Overview

Guinea-Bissau is situated on the West African coast (latitude 10° 59′ N, between 13° 38′ and 16° 43′ W meridians). It has a surface area of 36,125 km^2, maximum latitude of 193 km, and a maximum longitude of 330 km. The country has common borders with Senegal to the north and Guinea–Conakry to the south and east, and faces the Atlantic Ocean to the west. Its territory is divided into a continental zone and an insular one, the latter being composed of a contiguous chain of islands—Jeta, Pecixe, Areias, Caiar, Como e Melo and the Bijagós archipelago, made up of 88 islands and islets of which only 21 are inhabited. Since gaining independence from Portugal in 1974, Guinea-Bissau has been subject to considerable political instability. This political climate combined with mismanagement has contributed to a poor economy, and Guinea-Bissau holding the dubious distinction of being one of the most impoverished countries in the world.

2 Health and Welfare

The Republic of Guinea-Bissau is a developing sub-Saharan African country. Since its independence in 1974, there has been considerable socioeconomic instability, leading to disastrous consequences for the population's well-being, including poor access to basic healthcare. Recent data available at the World Bank database reveals that Guinea-Bissau has one of the lowest levels of per capita gross domestic product in the world (CINTESIS). Human development in Guinea-Bissau continues to be weak and precarious. This is gauged by the Human Development Index (HDI) which is a summary measure for assessing progress in three basic dimensions of human development: a long and healthy life, access to knowledge, and a decent standard of living. A long and healthy life is measured by life expectancy at birth. Knowledge level is measured by mean years of education among the adult population, which is the average number of years of education received in a life-time by people aged 25 years and older; and access to learning and knowledge by expected years of schooling for children of school-entry age, which is the total number of years of schooling a child of school-entry age can expect to receive if prevailing patterns of age-specific enrolment rates stay the same throughout the child's life. The standard of living is measured by gross national income (GNI) per capita expressed in constant 2011 international dollars converted using purchasing power parity (PPP) conversion rates. Guinea-Bissau has one of the lowest Human Development Index scores (0.420) ranking 178 out of 188 countries and territories in 2015. Between 2005 and 2015, Guinea-Bissau's HDI value increased from 0.388 to 0.424, an increase of 9.2%.

The two factors that contribute to Guinea-Bissau's low HDI are: widespread poverty with very low monetary income and limited life expectancy resulting from the lack of income-generating opportunities and access to quality healthcare.

Despite all the investment that has been made, overall data regarding maternal and child health worldwide reveal a serious situation. Maternal, neonatal, infant and under-5 mortality in this country far exceed the global average. Not surprisingly, the Guinea-Bissau birth and fertility rates are double the global rates. Furthermore, existing health resources are tremendously scarce, particularly human resources.

About 8.1% of GDP in this country is spent on health expenditures. Health conditions in Guinea-Bissau are among the worst in the world. Many people still suffer from such diseases as tuberculosis, whooping cough, typhoid fever, bacillary dysentery, and malaria. The poor state of health is perhaps best reflected in the country's high infant mortality rate of 101.64 deaths per 1000 live births, and a life expectancy of only 48.7 years of age. It was estimated that up to 20,000 people were living with HIV/AIDs in recent years.

3 The City of Farim and the Missionary Oblates in Guinea-Bissau

The first Missionaries Oblates of Mary Immaculate came in the seventeenth century to Guinea-Bissau, arriving at Farim on the border with Senegal. The Farim Mission serves two towns, Mansaba 30 km to the south and Bigene 40 km to the west, and a few villages.

Farim is located along the course of the Rio Cacheu, which is navigable up to this city, founded in 1641. According to an estimation in 2008, it has a population of 6405 inhabitants. The local population is dedicated to breeding and fishing and horticulture, and there are many women during the dry season who make salt harvest as their primary occupation. In fact, although it is located beyond 100 km from the Atlantic coast, the Farim area is an important center for salt productions and culture: the long stretch of the river Cacheu going from Farim to the ocean is actually an arm of the sea that creeps deep inland, a basin of brackish water subject to the alternation of the tides. This area is fighting against child malnutrition and neonatal and infant mortality remains extremely high (138 and 233 deaths per 1000 born alive).

In the context described in the previous paragraphs, the mission has focused over the years on these huge issues.

Firstly, with the design and then construction of Borboleta and Papagaio: respectively, emergency unit and Children's nutritional center. Both of the projects are presented in detail in the following paragraphs (Fig. 1).

Fig. 1 Farim village market of the local salt

4 Environmental and Climatic Analysis

Despite its limited territorial extension, Guinea-Bissau presents a vast variety of natural environments. Along the coastal plains a dense marsh of mangroves extends, which also dates back to the river estuaries hiding the banks under an intricate network of roots and aquatic plants. Far from the coast, the territory is covered by savannah, with grassy plains interrupted by acacia parasols, bamboo, palm, and banana trees: most of the land has been transformed into plots cultivated with rice, peanuts, corn, and palm oil. Farim has a tropical wet and dry savanna climate (Köppen–Geiger classification: Aw) with a pronounced dry season in the low-sun months, no cold season, and the wet season is in the high-sun months. The annual average temperature is 26.5 °C with an average monthly temperature vary by 2.5 °C: this indicates that the continentality type is hyperoceanic, subtype extremely hyperoceanic (Figs. 2 and 3).

5 Borboleta

The Borboleta infirmary (butterfly in Portuguese) is a small steel butterfly, a closed building structure to host children suffering from glaucoma in a space protected from the risk of infections causing blindness. Borboleta—commissioned by the "Gruppo 29 Maggio" Association NGO and supported by the Lecco Campus of the Politecnico di Milano—has also received a contribution from the construction industry, which provided materials and know-how for free (Figs. 4 and 5).

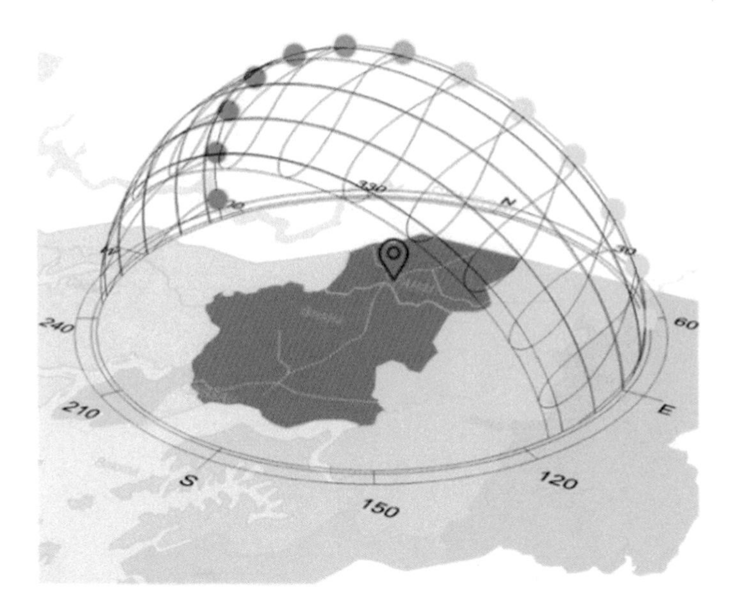

Fig. 2 Farim climate analysis: solar path

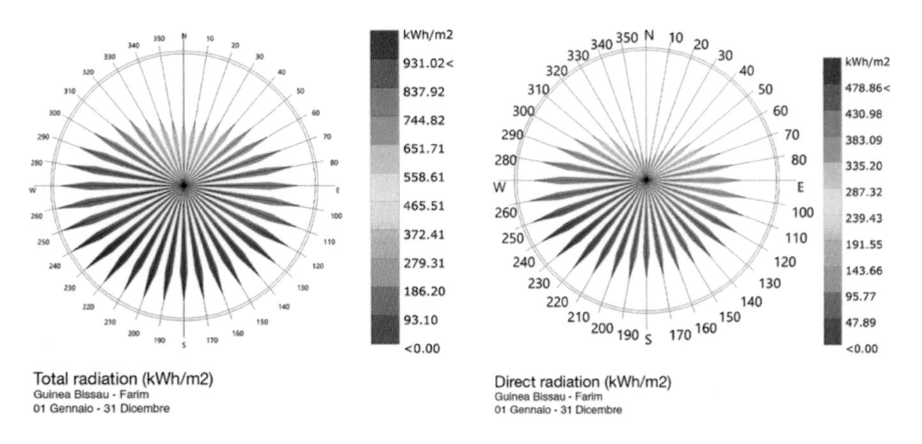

Fig. 3 Farim climate analysis: total, direct, and diffuse radiation

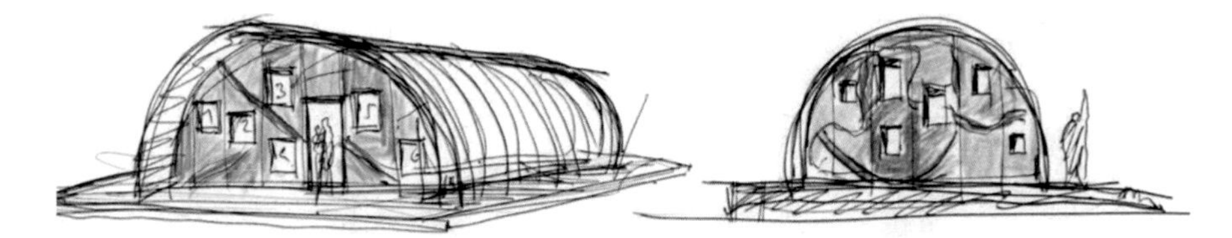

Fig. 4 First sketch of Borboleta (M. Imperadori)

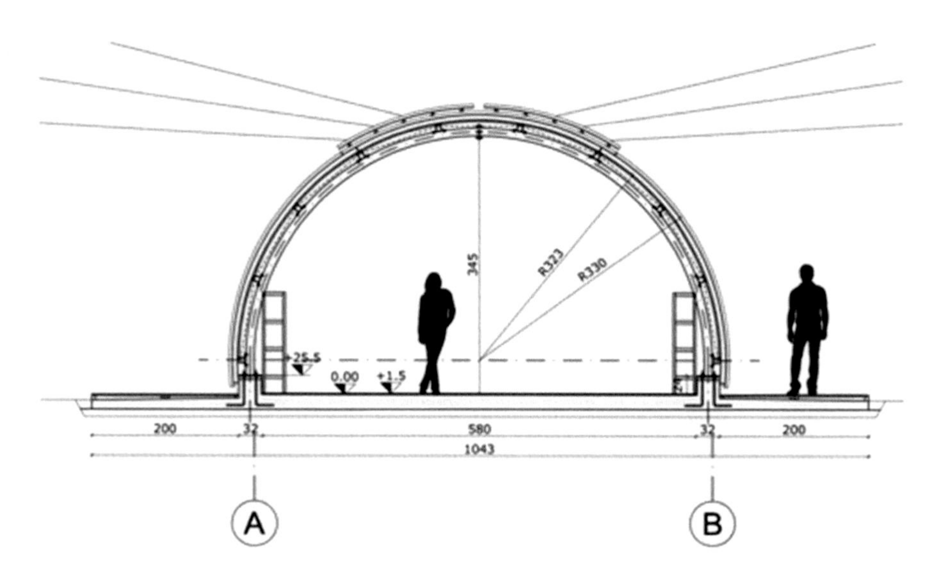

Fig. 5 Borboleta section

A key focus during the design phase has been dedicated to daylight analysis with the aim to increase the users' comfort. The daylight factor is guaranteed by two fronts, openings into the building envelope. The system plant foresees an air-handling unit that supplies two internal units and it has been designed to be powered by PV panels, which could be installed on the hull or, alternatively, on the surrounding buildings (Figs. 6 and 7).

Borboleta is a hull structure, which covers an area of 50 m^2. The deck is supported by a steel structure with three-hinged arches that carry cross-braced tubular braces in steel. The whole structure is hot-dip galvanized, protected by a triplex nanoceramic treatment, on which a powder coloring has been then applied. A recycled teak floor has been realized with a double function of waiting and meeting place for the patient's parents. The project includes the installation on the roof of shading nets, like butterfly wings, with tie-rod connected to the surrounding trees (Figs. 8 and 9).

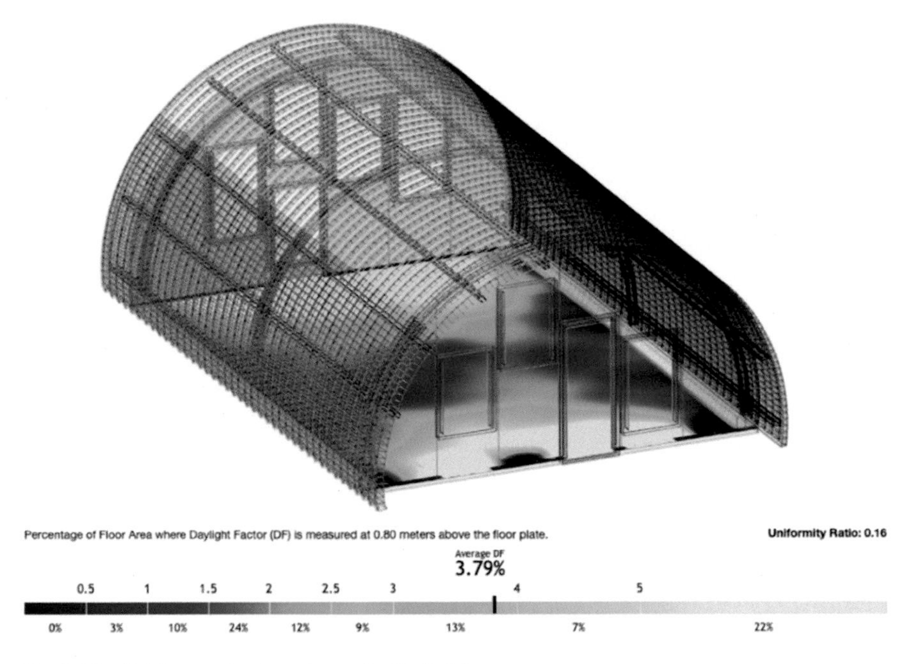

Fig. 6 Daylighting analysis performed with Sefaira

6 Papagaio: A Mixture of Commitment, Know-How and Solidarity

The Papagaio nutritional center was created in response to the need for a space to be able to exploit the food resources present on site and to fight child malnutrition. The realization was possible thanks to the partnership between the Politecnico di Milano, Carlo Pesenti Foundation, and sponsoring companies. The building was initially conceived as a block of 5 × 9 m with an entrance patio. During the construction site, due to material availability and the need of a dedicated distribution space, a building extension was realized (Fig. 10). The building structure is composed of ScaffSystem steel frame profiles, cold-shaped, coupled with multilayer insulated panels. The protruding roof shape, which gives the name to the building—recalling the Parrot wings, aims to protect the masonry from high solar radiation and from heavy rainfall in the rainy season. The earth block system wall, realized by local trained workers, allows for an increase in interior comfort. In addition, the small openings shielded by the roof overhang guarantee adequate illumination and respect for comfort even during daylight hours. The structure was assembled by local workers and a group of volunteers in less than a week, then the roof was connected and finally, the earth block system wall was realized (Figs. 11 and 12).

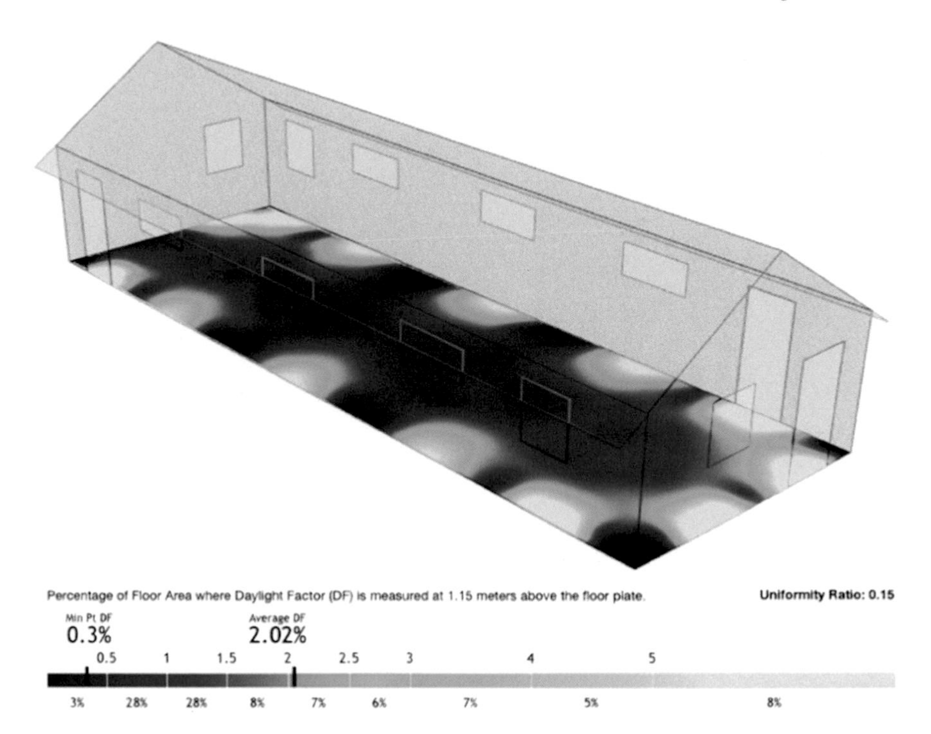

Fig. 7 Daylight factor analysis performed with Sefaira

Fig. 8 Borboleta construction phases: structures, non-load-bearing wall, finishing, and interiors

Fig. 9 Materials selection, transportation, and preparation for the construction site

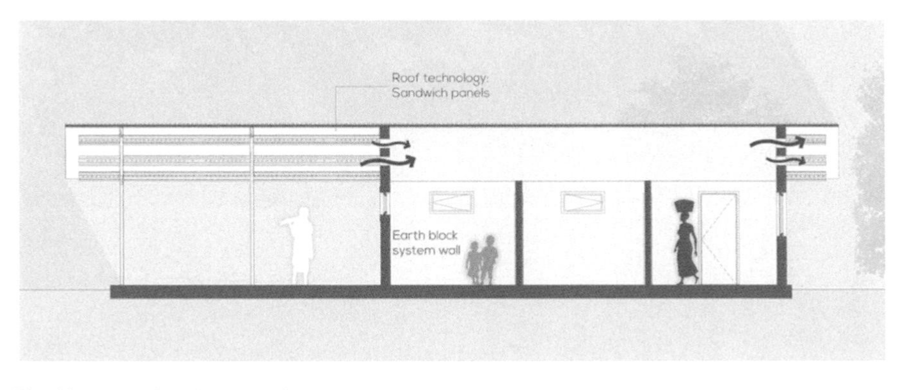

Fig. 10 Papagaio scheme design

7 Conclusions

In Farim's mission, it all began with a "limit-situation": many children died of malnutrition and others were abandoned by their mothers who couldn't provide for them. Borboleta and Papagaio were conceived to respond to those needs and over time became an ever more key sign of hope for the whole region: it hosted severely malnourished and recovering children, as well as any other sick children requiring pediatric care, with their mothers. Papagaio nutritional center has been dedicated to Franca Natta Pesenti of the Pesenti Foundation and Vincenzo Tamborrino, the visionary founder of Scaffsystem. Both of the buildings represent not only concrete results of years of experiences on sustainable built construction studies carried out by the Politecnico di Milano team but also the outcomes of fruitful collaborations between university and industries of the sector. Papagaio has been realized thanks to: the Politecnico di Milano, Scaffsystem, Carlo Pesenti Foundation, Mercegaglia, 3wood, Todeschini construction, Turra Meccanica, SFS, Emilgroup, Office Tamborrino and Kapriol Italia design.

Fig. 11 Papagaio construction phases: structures earth block system and name plaque fixing

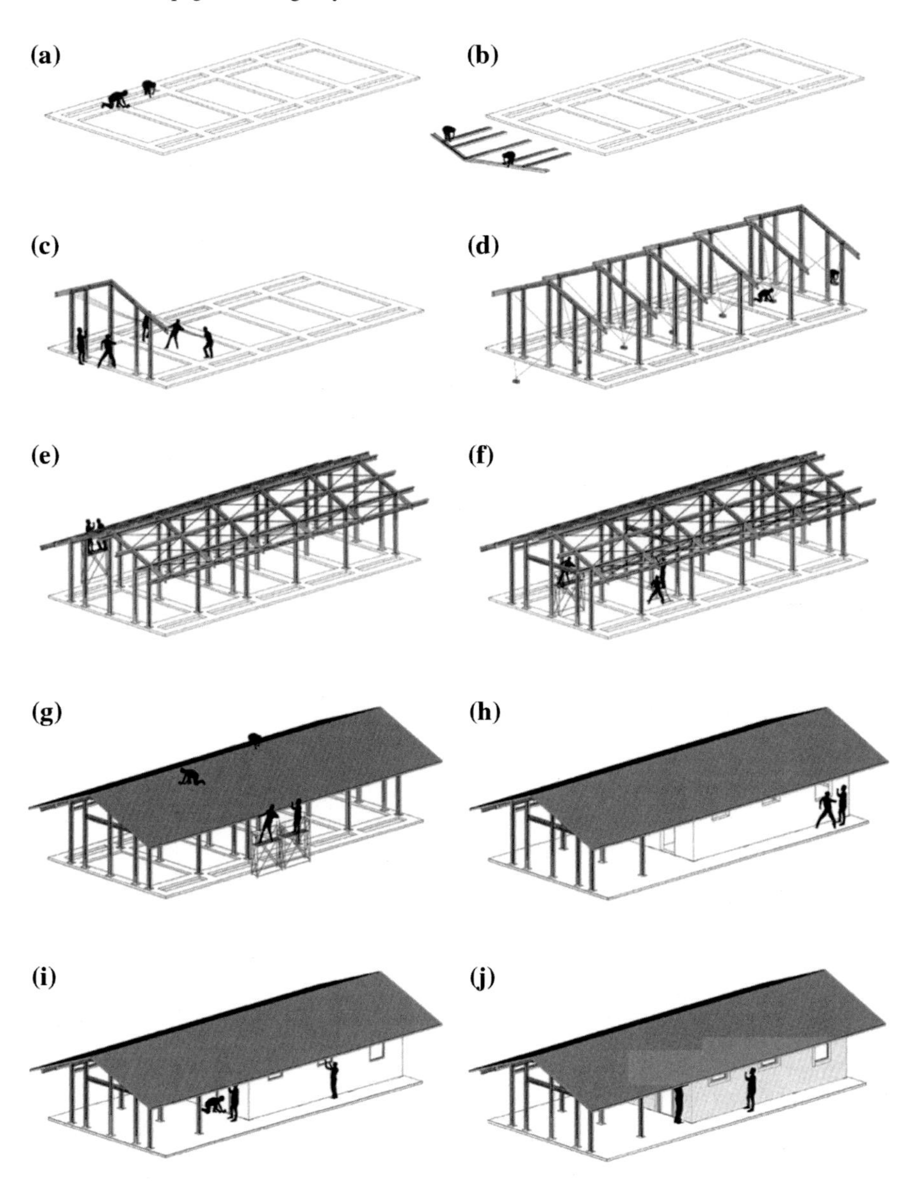

Fig. 12 Construction phases: **a** foundations, **b** frame ground assembly, **c** vertical fixing of the frame, **d** final fixing of all frames, **e** positioning and fixing beams and bracing on the roof, **f** positioning and fixing of beams and wall bracing, **g** positioning sandwich roofing panels, **h** earth system wall construction, **i** windows fixing, **l** plaster finishing

References

Adebajo A (2002) Building peace in West Africa: Liberia, Sierra Leone and Guinea Bissau. Lynne Rienner

Aruffo A (2003) L'Africa subsahariana: stati, etnie, guerre a sud del Sahara. Datanews Editrice

Butera M (2014) Sustainable building design for tropical climates. Principles and applications for eastern Africa. UN-Habitat

Centro de Investigação em Tecnologias e Serviços de Saúde (CINTESIS). Departamento de Ginecologia-Obstetrícia e Pediatria, Faculdade de Medicina, Universidade do Porto, Portugal

Coffman P (2015) West African tribal sculptures, textiles and artifacts study guide. American Alliance of Museums

Crivellaro A (2016) Africa e dintorni: scritture. Biblioteca dei Leoni

Forrest JB (1992) Guinea Bissau: power, conflict and renewal in a West African Nation. Westview Press BOULDER

Imperadori M, Salvalai G, Pusceddu C (2014) Air Shelter house technology and its application to shelter units: the case of Scaffold House and Cardboard Shelter installations. Procedia Econ Finan 18:552–559

Jammal M, Villa G (2008) Guinea Bissau: camminando nella palude. FBE

Mezzana D, Quaranta G (2005) Società Africane: l'africa subsahariana tra immagine e realtà. Zelig editore

Reynolds A, Zaky A, Moreira-Barros J, Bernardes J (2017) Building a maternal and newborn care training programme for health-care professionals in Guinea—Bissau. Acta Med Port 30(10):734–741. https://doi.org/10.20344/amp.8453

Salvalai G, Imperadori M, Scaccabarozzi D, Pusceddu C (2015) Thermal performance measurement and application of a multilayer insulator for emergency architecture. Appl Therm Eng 82(5):110–119

Novel Textile-Based Solutions of Emergency Shelters: Case Studies and Field Tests of S(P)EEDKITS Project

Alessandra Zanelli, Andrea Campioli, Carol Monticelli, Salvatore Viscuso and Gianluca Giabardo

Abstract The design of the first aid and reconstruction kits in major disaster scenarios highlights a new frontier for Technology of Architecture and Lightweight construction expertise, combining the traditional vocation of component design with innovative research on technical textiles. After an overview of S(P)EEDKITS—a project which received funding from the EU FP7—the essay focuses on research activities conducted by the ABC Department at Politecnico di Milano on the sustainable design applied to textile-based shelters. The essay also presents the results of two field tests conducted in Burkina Faso (2013–2014) and in Senegal (2015–on going), evaluating the environmental performances of shelters developed by the authors. The research was conducted in collaboration with International Federation of Red Cross and Red Crescent Societies, Netherlands Red Cross, the Médecins sans Frontieres (Operational centre of Amsterdam),—Operational centre of Amsterdam, Sioen Industries NV, Ferrino SpA, and with researchers of Vrije Universiteit of Brussel and Eindhoven Technische Universiteit.

Keywords Emergency · Shelter design · Packaging design · Textile architecture · Lightweight construction · Product development · Adaptability

1 Introduction

"S(P)EEDKITS: rapid deployable kits as seed for self-recovery" has been a collaborative project granted from the EU 7th Framework Programme; the call to which the project responded dealed with the rapid deployment of shelters, facilities and medical resources after a disaster. As specified by the title of this project, the objective was the development of novel emergency solutions, able to speed-up the response of humanitarian organisations (NGOs) during the first days after the disaster. Solutions

A. Zanelli · A. Campioli · C. Monticelli · S. Viscuso (✉)
Architecture, Built Environment and Construction Engineering—ABC Department, Politecnico di Milano, Milan, Italy
e-mail: salvatore.viscuso@polimi.it

G. Giabardo
Department of Design, Aalto University School of Arts, Design and Architecture, Espoo, Finland

N. Aste et al. (eds.), *Innovative Models for Sustainable Development in Emerging African Countries*, Research for Development, https://doi.org/10.1007/978-3-030-33323-2_10

needed to be clever and durable enough so that the affected population can (re)use them during the reconstruction phase. This dual approach—"speed" and "seed"—was crucial as the recent trend in emergency aid for NGOs is to stimulate as early as possible the affected population.

The S(P)EEDKITS project aimed to scrutinize the current used materials and equipment of the organization's Emergency Response Units (ERUs), and to develop novel solutions which drastically reduce their deployment time, the volume and weight for transportation. Normally, NGOs have sleeping ERUs in strategic world-wide warehouses, in order to deliver them immediately after the disaster strikes. Each ERU has specific modules (e.g. sheltering, sanitation, medical care, drinking water supply, basic energy needs) that can be used according to the assessment of the needs and that is intended for a specific number of people. Thus, the purpose of S(P)EEDKITS was to create new ERUs modules/kits that can contribute to saving the lives of people in the first days while planting the seeds for rebuilding future. One strategic key-point of this project was the use of technical textiles and lightweight construction to re-design and re-engineer current shelters and facilities of the emergency sector, thus developing more innovative solution that were easy to be handled, provided and transported.

The European Research Council (ERC) interest on the lightweight textile structures has been clearly evinced by the sub-topic of "lightweight construction, textile technology" under the "PE8: Products and process engineering" strategic research sector. That topic has been encouraging research and experimentation of new building solutions that go beyond the current accepted practices and standard applications of traditional materials to achieve both energy efficiency and structural stability through new lightweight materials, in particular technical textiles.

To reach these ambitious goals, the S(P)EEDKITS project team consisted of carefully selected partners, organized in workpackages (WPs). It was coordinated by the Belgian textile research centre (CentexBel) and also guided by the operational humanitarian actors like the International Federation of Red Cross and Red Crescent Societies (IFRC), the Netherlands Red Cross (NRC), the Médecins sans Frontières (Operational centre of Amsterdam) and the Norwegian Refugee Council (NRC).

The research activities were also supported by humanitarian research entities like Waste, Practica Foundation and the Internationales Biogas und Bioenergie Kompetenzzentrum (IBBK), while further key partners—like Milson B.V., De Mobiele Fabriek B.V., D'Appolonia SpA, Ferrino SpA and Sioen Industries NV shared their industrial expertise.

Furthermore, three academic partners were also involved—the Vrije Universiteit of Brussel (VUB), the Technische Universiteit of Eindhoven (TUE) and the Politecnico di Milano (POLIMI). They contributed their knowledge in the designing of new shelters, testing their performances through experimental field activities and optimizing their packaging and logistic aspects, with respective focuses on structural engineering (VUB), building physics (TUE), industrial design and architecture (POLIMI).

2 Research Methodology

Following a comparative analysis of terminology in different organizations—such as the European Commission's Civil Protection and Humanitarian Aid Operations (ECHO), the United Nations Office for the Coordination of Humanitarian Affairs (OCHA) and the International Federation of Red Cross and Red Crescent Societies (IFRC)—partners indentified the following common phases in the emergency management: emergency phase (first 48 h < time), relief (48 h < time < 2/4 weeks), recovery and development (time > 2/4 weeks). Before approaching the design of new kits dealing both with "speed" (in terms of transport/installation) and with "seed" (in the reconstruction phase subsequent to the emergency), it was important to investigate concepts of "kit" and "systemic design" in the traditional building process.

According to the Construction Products Directive—89/106/EEC, a "design system" can give rise to one or more kits, each of which may have different combinations of components. Following CPD directive (Fig. 1a), a construction product is a kit when it is a set of at least two separate components that need to be put together to be installed permanently (i.e. to become an "assembled system"). There are two possible types of kit those in which the number and type of components are pre-defined and remain constant and those in which the number, the type and the arrangement of components change according to a specific application. Some kits may be made up of one of many different possible combinations of components from a "design system" depending on the building in which it will be installed. Harmonized specifications shall cover kits in which the number and type of components are pre-defined and remain constant. They shall also cover an entire "design system", i.e. kits where the number, the type and the arrangement of components change according to a specific application.

In that sense, the S(P)EEDKITS design process was conceived as a "systemic design" on which all ERU kits are developed (Fig. 1b). The re-usability of some components of the kits was thus the peculiarity of this research approach, which aimed to create durable kits and reusable parts; each new kit can be different and customized,

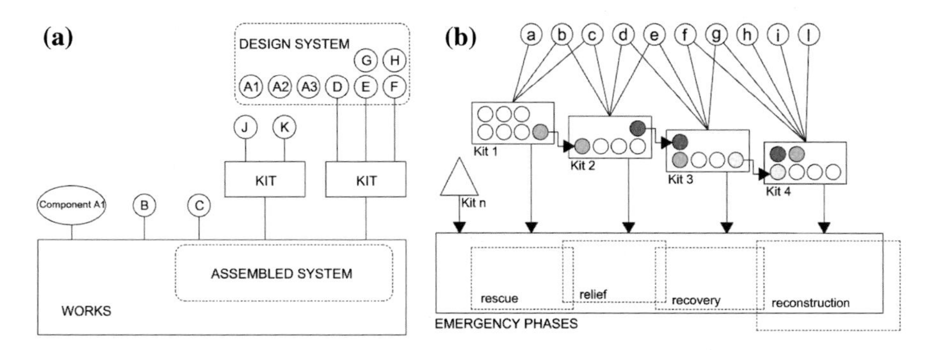

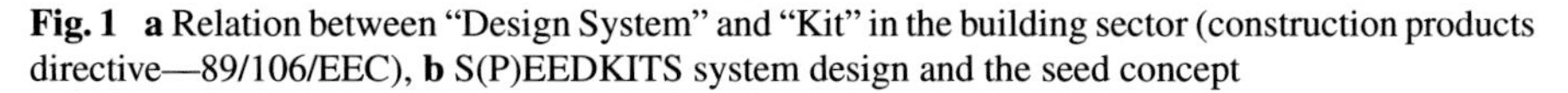
Fig. 1 **a** Relation between "Design System" and "Kit" in the building sector (construction products directive—89/106/EEC), **b** S(P)EEDKITS system design and the seed concept

targeted for a specific emergency phase, but with the following common peculiarities: (i) design system allows different combination of components in custom kits (for different climatic contexts; different cultures; means od transportation); (ii) some components can be reused in the subsequent emergency phases as a component of the forthcoming kit, which are later delivered; (iii) components should be ready from the very beginning relief phases but should later became useful building elements during the reconstruction phases.

As leader of "WP1—Modularity and packaging" and strategic partner of "WP2—Shelters", the role of POLIMI was mainly to find those "design harmonized specifications" which can be considered as a constant for many kits, in order to support each partner in systemizing relations between the different components of a kit to methodically plan the multiplicity of their "assembled system" which they should foresee to deliver in several emergency, climatic and cultural contexts and throughout different response phases.

The complexity of this research approach arised from the fact that a design system itself is made up of multiple nodes with multiple sub-nodes/sub-systems, with variable integration levels and transversal compatibilities among the nodes of this system. Even referring to Meadows (2008), designing a system (of systems) is about defining the relations between the different parts in an artifact within a boundary. The challenge for the POLIMI team was to determine those special emergent properties that make the kits developed in the whole project reciprocally harmonious and understandable for all users throughout the emergency process. Properties considered as emerging are listed below: (i) resilience, the ability to be robust and long lasting throughout different loops of uses; (ii) self-organization, that is the capability to self transforming starting from a modular scheme to a more complex shape; (iii) hierarchy of the whole system, the associative rules based on a bottom-up evolutionary approach. More specifically, requirements followed in the development of new shelter concepts and packaging framework were dealing with the concepts of lightness, ease of transport, low price, simplicity, modularity, ergonomics, reliability, robustness, durability and multiplicity.

3 Packaging Design

At the beginning of the research activities of "WP1—Modularity and packaging", the POLIMI team explored and evaluated several commercialized packaging solutions that through a transformation can change their shape, volume or function. The state of the art (SOTA) mapped some interesting solutions, organized into two cross-matrixes: forms/materials and transformational dynamics/materials. The first map considered different shapes and lays them out on a matrix with various materials usually adopted in packaging solutions; the second one outlined different dynamics that can happen within the packaging itself to have it change its shape, function or volume (Zanelli et al. 2014).

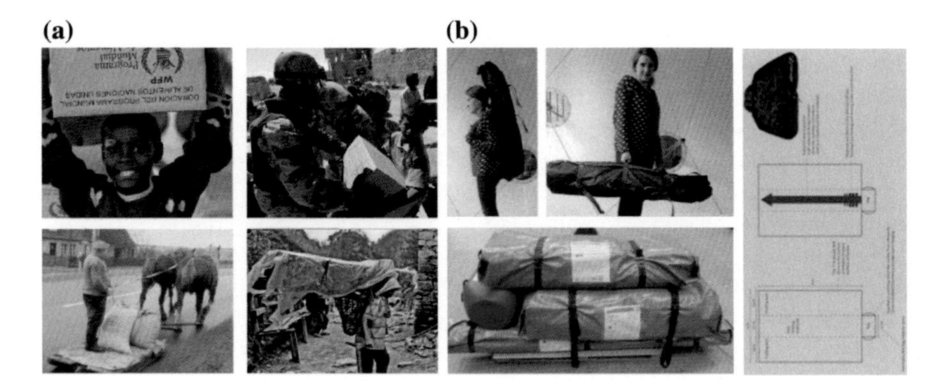

Fig. 2 **a** Current packaging in emergency field, **b** S(P)EEDKITS basic bag concept

In order to list profitable examples and inspiring solutions that can be transfer from other sectors to the emergency field, special attention was done to the following design principles, thus qualifying the state of art as a systemic catalogue of multipurpose concepts: (i) the rhizomes as system of systems; (ii) the nesting as an optimizing process; (iii) the identity and information layer; (iv) the handling as the key strategy for new smart packs made of textiles. Taking into account these design principles, POLIMI worked on a novel packaging concept ("basic bag"), used for WP2 shelter kits, and on the optimization strategies that all the partner—involved in the consortium as kit developers—needed to apply. Strategies were targeted at maximizing occupation, managing modular composition of items in the load, reducing voids and minimizing handling (Fig. 2).

The developed "S(P)EEDKITS basic bag", considered as a pilot concept design, is flexible enough to become the founding element upon which it is possible to develop the modular, rhizomatic system. The basic bag concept consists in a module made of durable technical textiles, produced in two diverse sizes derived from a subset of the Euro-pallet (length = 120 cm; width = 80 cm). Through a cyclical replication of the sizes, it is possible to obtain a maximum of six combinations that satisfy all the dimensioning variations expected in S(P)EEDKITS. Referring to the bag material, the research focused on layered HDPE/Kraft Paper, non-woven textiles, PP and PE woven textiles. The final design also included material labelling with information of further potential uses for the end of life/recycling.

3.1 *The Rhizomatic Concept*

In 1983 the philosopher Gilles Deleuze and the psychoanalyst Félix Guattari used the term "rhizome" and "rhizomatic" to describe theory and research that allows for multiple, non-hierarchical entry and exit points in data representation and interpretation.

In “A Thousand Plateaus” they opposed it to an arborescent conception of knowledge, which works with dualist categories and binary choices. A rhizome works with planar and trans-species connections, while an arborescent model works with vertical and linear connections (Deleuze and Guattari 2004). Transferring this theory into the packaging design, POLIMI team noted divers suggestions of on-market products in which all packed elements were horizontally compatible in terms of stacking, loading, transporting and assembling. After analysed this peculiarity in SOTA, the ideal output in S(P)EEDKITS was a coherent and homogenous family of packaging that can always interact with each other at different levels: size-compatible, volume-optimized, modular, light, easy to handle, easy to recognize, understandable in terms of communication and instructions, compatible in terms of identity.

3.2 The Nesting Concept

In packaging design, the nesting concept approaches the issue of optimizing volume occupation and weight, possibly generating modular elements that can work on any level, from the bag/box to the pallet and the container levels. The development of a nesting approach was the core of the POLIMI research activity on WP1, focused on making the process of packaging and the logistic chain simpler. A relevant part was the dimension/quantity constraints of the basic element or component at the bag level, that can reproduce itself or multiplicate and be accommodated in the pallet/container level.

3.3 The Visual Identity Concept

The identity concept deals with the way packaging and its content are identified and identifiable. This area of the project intervenes on a layered set of instances, from the system identity on a “brand” level to the functional one. In S(P)EEDKITS, the goal of the identity concept was thus to integrate an “across the board” tagging/labelling/information/instruction/branding communication system for the whole research project. A specific research into the colour coding and branding was conducted with the NGOs—involved into the research project—that are active in the field (namely IFRC, NLRC, NRC). Furthermore, a visit was organized to the German Red Cross (GRC) Logistic Centre in Berlin to further explore and experience the matter of packaging, stock management, logistic and markings. POLIMI studied the way GRC labels their equipment. They use a mixed colour/text labelling that refers to a coding system that is well known within the organization but that might, however, not work so well with “outsiders” who are not familiar with the markings and abbreviations. Starting from those assumptions, the POLIMI team worked on the development of a layered system, where S(P)EEDKITS identity is clearly communicated and always present together with all the contents related to tagging–labelling–information–instruction, while keeping an open possibility for customization for the acquiring organization. The research team also explored ways to render accessible the colour code to people suffering from colour blindness.

3.4 Handling as the Key Strategy

As S(P)EEDKITS was focused on lightweight solution easy to transport, partners were induced to develop several bag-based kits, like the shelters of the WP2, with a keen eye on the resistance, durability and protectiveness of the packaging. To optimize bag/box useability (third packaging level) POLIMI analysed different ways of human-handling the objects. This task was strongly linked to habits, cultures and traditions and their differences. The three main ways that were mapped are: (i) lifting; (ii) rolling/sledging; (iii) transporting on-body (head, shoulders etc.). This overview gave a panorama of the different conditions of packaging that the POLIMI team followed in order to find easy ways of handling and transporting the kits while minimizing the inherent risks connected to the transport and use.

4 Shelter Design

In humanitarian field, shelter products are mainly developed as "closed" prefab systems that work independently to other provided shelters and local materials. Prefabricated designs are developed ad hoc and their parts often require time-consuming assembling. Sometimes prefab products do not include instructions for post-emergency use or disposal. As result, abandoned temporary shelters become common, sad reminders of the easy waste of money and resources. Moreover, the different climatic contexts require from NGOs a huge faculty of adaptation as each situation calls for a precise answer. Recent emergencies draw attention to limits of current standard tent to be adapted in all climates or in places with high daily-temperature ranges (Aquilino 2011).

For overcoming this critical aspect of current shelter kits, the development of novel solutions aimed to offer an effective winterized solution that also well works in warm and hot climates. The idea of a progressive solution was adopted according to local constraints: it was not only linked to climate risk, but also dependent by local resources (Virgo et al. 2014). Adaptability was to ensure both a prompt first-time repair, that can be easily erected, and an effective protection in a medium and long-term period, so configuring the "core" of a transitional dwelling.

Moreover, a novel shelter system should not only link to climate risks and local resources, but also relate to cultural identity of the affected population. The novel shelter kit has to be inserted in an affected area (urban area, improvised camp, rural region etc.) to reach as quickly as possible an acceptable post-disaster situation towards the rebuilding of economic and social life. By providing shelter kits that are adaptable to users' practices (tribal composition, lifestyle, religious claims, etc.), the rescue could be organized with a people-centred approach in which refugees enclose themselves private spaces, even inside damaged buildings. This feature can improve the acceptance level of the entire sheltering process during a disaster.

The concept generation aimed to transfer the practical users' needs in a set of shelter concepts. In extreme climate conditions, the first basic need is to protect the displaced population against external agents. On these fields, the lack of insulation of standard tents and damaged building is more critical than the request of structural elements (timber or steel) and blocks, which are easily recoverable from local markets. Thus, potential novel shelters should mainly provide flexible panelling systems to adequately winterize beneficiaries; they might be compatible with standard tent structures and locally available structures (frames, simple poles, trees, etc.); finally, they could be adaptable to diverse functions depending on needs (e.g. the roofs after a hurricane, floors during flooding, etc.).

Requirements and inputs coming from S(P)EEDKITS partners (IFRC, SIOEN and CENTEXBEL) were translated in a list of metrics. The list reflected as directly as possible the degree that new concepts had to approximate for satisfying emergency needs and production skills. As sketched in Fig. 4, in product design, the workflow usually starts establishing a set of specifications, which spells out in precise, measurable detail what the product has to do (Ulrich and Eppinger 2011).

4.1 Textile Wall

The first concept developed by POLIMI referred to a tri-dimensional cell panel (named "textile wall") that can deliver on roll and configure diverse layout on the field. By using the roll panels to enclose a living space, one or more tarpaulins can cover wall layouts, while the poles used to tension roof membrane also stiffen the corners of panel. After an initial use as simple membrane partitions (with cell empty or partially filled for ballasting the base), panels can be gradually stuffed or used as formworks for concrete. For this aim, it could encourage local users to reuse panels to repair or rebuild homes, taking into account own constructive and social background. Panels should also work to construct raised basement or to cover flat or pitched roof, with cells used like air cavities or insulated (Zanelli and Viscuso 2015).

On October 2013, SIOEN fabricated 12 m-length prototype of textile wall with thickness of 20 cm and height of 150 cm. The roll was shipped in the refugee camp of Sanioniogo (Burkina Faso). From the 3rd to the 10th November, IFRC and POLIMI set-up the wall under a tensioned roof (developed by VUB University of Bruxelles), with the support of Lux Red Cross and Burkina Red Cross (Fig. 3). In that case, due to the dry environment (and soil) it was not possible either to prepare trenches

Fig. 3 S(P)EEDKITS field test in Sanioniogo, Burkina Faso, November 2013

or to collect a large amount of filling material. Therefore, it was not possible to fill the panel without inserting structural elements into its small cells. After positioning the Textile wall along a covered area of 15 m^2, locally available dried eucalyptus poles were used to prepare two frames made with horizontal and vertical elements crossing each other every 1 m approx. Connections were fixed with ropes. This first prototype erected in Sanioniogo has been shown during the MILIPOL, the annual exhibition about safety and risks, held in Paris from 17th to 20th November 2013, achieving a good evaluation from NGOs and field operators.

4.2 *Cocoon*

The second concept consisted in a complete living accommodation (dimensioned on module of 3.60 × 1.80 m), that can fix to whatever structural element by means of polyester belts. It allows crating a confined, winterized space to assure intimacy and protection. The amount of material needed oriented towards a lightweight insulating material, such as a non-woven polyester fabric with thickness of 20 mm.

On December 2015, partners shipped 10 cocoons to Senegal for a field test in Ntiagar, a village close to Dakar that is inhabited by Ronkh community (Fig. 4).

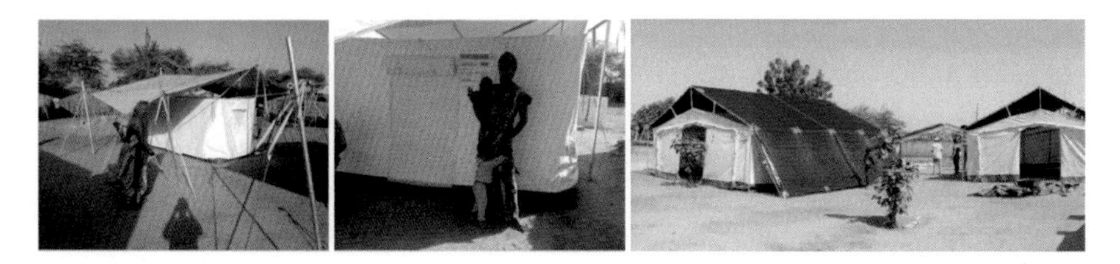

Fig. 4 S(P)EEDKITS field test in Ntiagar, Senegal, December 2015

Taking into account packaging guidelines developed within the WP1 modules were wrapped with thermoplastic film in a kit including eight polyester belts provided with tensioners, four pegs and the instruction manual. Once arrived on site, the weight of the whole bale (30 kg approx.) allowed a carry-on local transport (1 person per bag). The Luxemburg Red Cross used the modules together with poles provided within the tensioned roof developed by VUB. Products were observed during a three-month's period, characterized by low temperatures (that sometimes lowered to 12/12 °C during the night) and very strong winds (Harmattan winds) with forces often superior to 60 km/h and heavily charged with dust. This represented an important test element since the illnesses due to dust and wind in the winter Saharan season are widespread.

4.3 Multipurpose Tent

The multipurpose tent is an emergency shelter designed for hosting collective activities and functions that need bigger use surfaces (accommodations, medical room, meeting room, etc.). For that reason, particular attention was given to the usability of the inside space (e.g. vertical walls instead to inclined sides) that increase the net internal area and facilitate the connections between units (both on the front and along the sides), allowing the creation of bigger spaces and camp infrastructures. A single unit measures 48 m^2 and was designed to host between 10 and 12 people. All the connectors are the same but rotated, in order to facilitate the set-up in emergency contexts. They functions also as shade net antennas, allowing the elimination of additional framework, the shade net can be integrated in the basic kit. The fabric layer is composed by a breathable material (polycotton) for the upper part and a waterproof groundsheet (not separated) for the bottom layer; it is hanged from inside the structure in order to ease the assembling. Although the tent has a net internal area of 48 m^2, the structural design is compliant with snow and wind load specifications of UNI EN 13782:2015 (Temporary structures. Tents. Safety), that describes safety requirements which need to be observed at design, calculation, manufacture, installation, maintenance, of mobile, temporary installed tents with more than 50 m^2 ground area. S(P)EEDKITS partners observed functionality and reliability of two multipurpose tents during the test in Ntiagar (December 2015). During the test, NGOs and local people verified that the mounting is simple and intuitive, also because it can be set-up without the need of special tools (Fig. 4).

5 Conclusion

This contribution is intended to describe the design of novel packaging principles and shelter components that contrasts the global production of emergency items through a "glocal" approach. This feature may change the current, local perception of sheltering provision because it links prefab production and its standardized transportation with local markets and relative construction technologies, favouring the acceptance from beneficiaries (Charlesworth and Ahmed 2012). Within the S(P)EEDKITS project, the collaboration with NGOs and manufacturers allowed to include in the design process both the real needs on the field and the current technologies in sheltering production. Metrics coming from requirements permitted a selection between diverse design concepts; finally, the most effective one was prototyped and tested on the field.

Referring to POLIMI research outcomes, both packaging principles and shelter design achieved good evaluations from humanitarian operators that highlighted how the mounting is very simple and intuitive. In any field test organized within the dissemination activities of S(P)EEDKITS, after a quick demonstration of the mounting, local people were able to perform the mounting by themselves, without making mistakes. After two years from the field test in Senegal, shelters are still used by beneficiaries: they are in the same state as they were when mounted, without any sign of deterioration. Users also observed that the shelters are cleanable and maintainable in a good state. Internal temperatures are acceptable thanks to the shade nets during the day and the insulation has allowed maintaining a good temperature during the night. Moreover, mosquito net layers avoid the entrance of sand or dust inside the shelter notwithstanding Harmattan winds.

References

Aquilino MJ (2011) Beyond shelter: architecture and human dignity. Metropolis Book, New York

Charlesworth E, Ahmed I (2012) Shelter and disaster risk reduction in the Asia-Pacific region: final report. Humanitarian Architecture Research Bureau (HARB), RMIT University, Melbourne

Deleuze G, Guattari F (2004) A thousand plateaus: capitalism and schizophrenia. University of Minnesota Press, Minneapolis

Meadows DH (2008) Thinking in systems: a primer. Chelsea Green Publishing, White River Junction, VT

Ulrich K, Eppinger D (2011) Product design and development, 5th edn. McGraw Hill/Irwin, New York

Virgo V, De Vilder I, Viscuso S, Roekens J (2014) Field study Burkina Faso, S(P)EEDKITS WP2 report, 27 Jan 2014

Zanelli A, Viscuso S (2015) Flexible panel—Pannello flessibile. Application numbers: WO2016166658A1. Assignee: Politecnico di Milano

Zanelli A, Buyle G, Giabardo G, Viscuso S (2014) S(P)EEDKITS & smart packaging. Nuove applicazioni tessili per ridefinire la risposta alle emergenze. TECHNE, vol 8

BY

The Mo.N.G.U.E. Development and Experimentation Project in Mozambique

Liala Baiardi

Abstract The Development and Experimentation project in Mozambique led to the elaboration of a model and an investigation process aimed at delineating a cognitive base useful to trace possible improvement actions to be undertaken while respecting the preservation of the place's identity in accordance with developmental continuity. The project is located in a territorial context of great natural value which is being restored as a social and identity point of reference. For this reason, it needs protection from uncontrolled urban growth that could undermine the community-environment system. The general objective is to make Mongue a place with a recognisable identity for a large community through shared interdisciplinary planning aimed at establishing a plurality of functions and services. The aim is to create a new system capable of feeding that rooting in the place, necessary to counter the tendency to eradicate local populations.

Keywords Process approach · Development · Sustainability · Feasibility · Interdisciplinarity

1 Introduction

The transformation process which is going through Africa is also highlighted by the significant increase in the urbanisation rate, which has risen from 15% in 1960 to 40% in 2010 (UN-Habitat 2015). It can be seen that the population in Mozambique has increased from 5 to 25 million inhabitants in the last century. A strong indicator of this rapid growth is that 45% of the population in 2015 was younger than 14 years (Instituto Nacional de Estatística 2015). The significant increase in population has generated critical issues related to fast urbanisation, the depletion of environmental resources and the wealth of local cultures. The consequences are, for instance, poor quality living spaces and housing facilities, lack of places for social gatherings and

L. Baiardi (✉)
Architecture, Built Environment and Construction Engineering—ABC Department, Politecnico di Milano, Milan, Italy
e-mail: liala.baiardi@polimi.it

N. Aste et al. (eds.), *Innovative Models for Sustainable Development in Emerging African Countries*, Research for Development, https://doi.org/10.1007/978-3-030-33323-2_11

Fig. 1 Town of Maxixe (author's photo)

educational facilities for children and serious problems concerning the safety and hygiene of places and people.

As already highlighted by previous studies (Monga and Lin 2015; Scott 2015), it is considered impossible to overcome poverty or the growth challenge in Africa without considering the importance of urbanisation.

The poor construction quality of the buildings and the limited application of maintenance actions, together with the lack of spatial regulations to support the strategic planning of the territory's development, significantly reduce the capacity of the system to resist climatic variations, to adapt to the effects of disorder and to regenerate following change whilst preserving its functions and identities.

The development practice of the Mozambican territory is predominantly characterised by the lack of an interdisciplinary design methodology applied to the reading and the design of places, capable of enhancing the area with its natural and social landscapes, its historical and architectural heritage, space liveability and environmental sustainability aspects. In this context, the Mongue Development and Experimentation Project is proposed with the aim of helping to bridge the design methodology gap that characterises many interventions in these areas, which are often the result of the generous but sometimes confused spontaneity of international social actors or their lack of knowledge/experience (Fig. 1).

2 Territorial Framework

Between 1498 and 1975, Mozambique was a Portuguese colony. The country gained its freedom after a decade of guerrilla warfare (Thompson 2013). Only one year after its independence from Portugal, a rebel group called Resistência Nacional Moçambicana (RENAMO) opposed the liberation movement generating a series of internal uprisings that lasted until 1992. The number of victims of this internal struggle is estimated at between 600,000 and 1 million people (World Peace Foundation 2015). The country, which was already in a state of poverty, found itself in an even worse situation.

The situation has improved in recent years: according to a national survey, the number of Mozambicans living in absolute poverty has been reduced from 70% (in 1997) to 54% (in 2008–2009). However, most of the rural population still lives on less than $1.25 a day and lacks basic services such as safe access to water, sanitation and schooling (IFAD 2014).

Surveys carried out on the country's development between 2000 and 2005 (Ollivier and Giraud 2011) highlight that the wealth of Mozambique has increased mainly through the accumulation of human and physical capital, while pressure on renewable natural capital remains relatively low. Although in the past, there was an average annual growth of 7% in GDP (Gross Domestic Product), it is expected that by 2040, there will be 18.7 million people living in absolute poverty, which is almost the same as today (19 million) (Porter et al. 2018).

The rapid and uncontrolled urbanisation (IHS 2017) spreading across the country favours the creation of shanty towns: in 2009, around 80% of the urban population lived in impoverished districts (WHO 2014).

The government is actively fighting the country's main problems. Some examples include the WASH project (water, sanitation and hygiene) promoted by UNICEF for access to safe water, sanitation and hygiene, and the PARPA project 'Action Programme to Reduce Extreme Poverty' developed by the Ministry of Planning and Development of Mozambique (IMF 2011) (Figs. 2 and 3).

Fig. 2 Example of the harmful effects of uncontrolled urbanisation. Episodes of soil erosion due to torrential rains in a typical shanty town in the village of Maxixe (author's photos)

Fig. 3 Example of the harmful effects of uncontrolled urbanisation. Episodes of soil erosion due to torrential rains in a typical shanty town in the village of Maxixe (author's photos)

3 The Mo.N.G.U.E. Mozambique Project. Nature. Growth. University. Education

The heading, a play on words on the name of the town 'Mongue' (a town in Mozambique), aims to bring to light certain general objectives that are considered important for the reference context:

- The big theme of nature, the environment and landscape and its hazardous condition and potential;
- The unique theme of growth primarily understood as development in a qualitative and sustainable sense;
- The university as a major driving force of economic, but above all, civil development in the country;
- The grounds for widespread education, education for all, which now also extends to nursery.

The research proposal, selected from among the winners of the Politecnico di Milano 'Polisocial Award' (2016 edition),[1] relates to the local authorities' support in safeguarding the landscaping of the territory. It proposes the creation of a protected area by establishing the Ecological Park at the University of Pedagogy in Maxixe.

The research involves an application-oriented testing activity on a project to test a methodology replicable in similar contexts. The aim is not to define prefabricated modules that can be assembled and reassembled again in an undifferentiated manner,

[1]Politecnico di Milano working group: Liala Baiardi, Michele Ugolini, Valentina Dessi, Rossana Gabaglio, Laura Montedoro, Lorenza Petrini, Stefania Varvaro, Luca Faverio, Filippo Ganassini and Marco Talliani.

isolated from the context and environment in which they are placed, but rather to develop a useful method in a conscious design practice with new interventions: an interdisciplinary working path (mapping, analysis, strategies, methods, projects, etc.) capable of interpreting the demand for improving people's living conditions (quality and livability of spaces and the environment with a view to environmental sustainability).

4 Development Methodology

The research led to the development of a model and a process of inquiry aimed at outlining the state of affairs and suggesting possible improvement to be attained as regards preservation of the place's identity according to a developmental continuity.

The experimental aspect has allowed for the model to be refined (application phase) directly in the field, resulting in inspections being scheduled within the context of Mozambique as a reference point.

Mongue was identified as an area of study and experimentation during the application phase: it is a peninsula that stretches for about 20 km north of the town of Maxixe. The peninsula is of particular interest, as it contains certain rural context phenomena (spatial neglect, lack of care for the environment, etc…), and at the same time, it risks suffering the effects of the settlement tensions that characterise the peripheral areas of the urbanised cities. This place is also full of identity and history in that, in addition to the rare presence of buildings dating back to the period of the Portuguese colonisation, the first missionary home was built there in southern Mozambique and abandoned during the Revolution and the Civil War (1964–92) (Fig. 4).

Fig. 4 Mongue mission (author's photo)

The scheduling of the screening process resulted in the planning and processing phases of the work summarised as follows: the development of relations with public institutions, residents and local professionals; multidisciplinary thematic mapping and analytic interpretation; definition of a framework of levels and indicators to define qualitative and quantitative metrics; definition of an intervention strategy for the study area through an interdisciplinary design method.

The working hypothesis is necessarily based on multidisciplinary assumptions that define the different levels of the project in order to achieve a common goal (interdisciplinary). The areas of competence put in place to fulfil the various analytical and design aspects highlight the multidisciplinary nature of the project:

- Architectural, interior and open spaces, urban, regional and landscape;
- Preservation of historic buildings and existing buildings; structural-technical and bioclimatic, environmental and energy sustainability;
- Economic enhancement, financial sustainability (Monti and Romano 2010), management and building maintenance.

The added value of this multidisciplinary approach lies in making operational syntheses to overcome dystonias arising from mono-disciplinary approaches (the latter, in fact, have the limitation of not considering the complexity of the problems that affect the spaces of our lifestyles at all levels). This requires a careful approach to local resources/conditions that will define a framework of useful levels and indicators to clarify the qualitative and quantitative metrics identified as follows:

- Usage patterns of public and living spaces;
- Analysis of materials and traditional construction techniques;
- Analysis of the bioclimatic and environmental conditions;
- Analysis of local energy systems/water supply production;
- Acquisition of historical and socio-cultural knowledge.

The added value of the identified methodology is also given by the synergy between the skills of the Politecnico's applicants with the different local knowledge: advanced training from the Pedagogical University; the knowledge gained in the field by the missionaries from the Sagrada Familia; the needs expressed by local communities (Fig. 5).

5 Results

With the proposal to establish a park and a research centre, the aim is to sustainably design spaces adapted to territorial needs and local communities, by way of strengthening the services offered, creating a socialisation process aimed at physically transforming places, and making room for better living conditions.

Fig. 5 The water supply ritual. The Mongue area only has one single distribution point for drinking water, and there is no water distribution network to homes (author's photo)

Using an analytical-interpretative approach, focused on the specificity of the places and the social problems, the research project paid close attention to the natural, environmental, landscape, architectural and social resources in the area. This aims to bridge the design methodological gap that has characterised some interventions made by Western operators in the context of developing countries, sometimes with high architectural and performance qualities but substantially unrepeatable and which, beyond the intervention itself, do not leave any methodological seeds for future developments in the area (Fig. 6).

The research activity led to the development of a master plan which, in addition to the relationship between the buildings, also takes into appropriate consideration the system of liveable and vital open spaces, with the ability to attract the local community and create a sense of affiliation to the place. The next step concerns the building design, focusing in particular on the internal distribution of the premises in order to provide the best living conditions that can optimise the use of materials and energy resources available in the area and ensure the best hygiene and thermohygrometric comfort conditions.

'Data' on the effectiveness of the design are also evaluated through the use of quantitative indicators and qualitative parameters that were used to build a methodological analysis and therefore represent the project's fundamental footprint. The green system, the built system, the system of techniques and materials, the soil levels system, the landscape system, the connections system and the energy and bioclimatic systems must all find mutual verification. Some aspects are more easily verifiable through the use of commonly used quantitative indicators, especially for economic and energy-environmental aspects, while other aspects related to the themes of architectural space can only be measured quantitatively in a partial and somewhat inefficient way

Fig. 6 Project at the Research Centre in the Mongue Park. Authors: the MoNGUE working group

(square metres, volumes, regulatory ratios, etc.). These aspects find their verification potential in another manner through graphic diagrams of comparison between the factual state and the project, 3D simulations on the design of open spaces, green areas and buildings, including the various evaluated hypotheses, photo inserts and overlapping territorial sections. Data collected on site through instrumental measurements, photographs and physical reconnaissance-interpretations serve to demonstrate the appropriateness and in-depth study of the project. Verifiable aspects through the use of quantitative indicators:

– Verification of the enhancement: areas usable for group activities and socialising; service activities for children (number of children accepted into new school facilities, increasing the useful space for children); area recovered in the historical heritage of the mission and its corresponding enhancement; accommodation capacity (number of people housed in the accommodation).

- Project's socio-economic feasibility: number of jobs created (which helps to define the increase in financial support that local families can receive), Net Present Value (NPV, which is the discounting of all positive and negative cash flows that characterise the project), the project's Internal Rate of Return (IRR) which is useful for comparing the project's alternative hypotheses and the investment repayment periods (Payback period) representing the number of years required to repay the investment.
- Environmental performance: the evaluation of the parameters summarised in the project, linked to the orientation, the distribution of the environments, the placement of the openings and the building system, which regulate the availability of passive energy resources, is assessed through the use of a thermal comfort indicator (PMV, for example), and simultaneously the primary thermal energy demand required to meet thermal comfort conditions.
- Energy saving during the production of the materials, the construction and maintenance of buildings, for monitoring and comparing pre-variations and post-interventions with sizeable interest, such as traditional biomass savings, the efficiency of cooking systems, kWh products and those used through conventional and alternative systems, and fuel savings for the generators.

6 Possibility for Future Developments and Weaknesses

The expected results are to share the design methodology with local players and, thanks to close cooperation with the local university system, the training of professors and students along with a growing awareness of the local communities. Direction, in all its phases (planning, design, construction and programming), is crucial to the success of any initiative and is the fundamental basis of satisfactory management, including socio-economic, of the building and the territorial system over the years.

Adopting this approach, however, involves the effort of being able to associate the main features (uniqueness, physicality and immobility) of real estate with mathematical and financial elements.

Real estate is first and foremost a physical asset, closely related to its context, and presents unique characteristics that make it difficult to compare with other properties without using type approval parameters.

In addition to the local dimension affected by the intervention, the project has the potential to cause an impact on a larger scale, on the region or the country, where it will become known as best practice and will bring a dimension of complexity to the transformation initiative even in low-cost areas (Table 1).

Table 1 List of the main benefits arising from the project

Benefits deriving from the project success
Social
Restoration and enhancement of the mission facilities at disposition of local people: school, canteen and church
Creation of 17 new job positions during the whole year
Possible improvement of general conditions of the surrounding area
Improvement of the services present in the area
Possible improvement of the infrastructure system connecting Mongue with Maxixe
Introduction of an electricity system
Creation of a reference point for local population
Creation of an international educational meeting point
Creation of an exchange of information and resources network between universities
Collaboration with other facilities in the area or acquisition of external services from the community
Construction of a small dock that will connect Mongue directly to Inhambane and to their villages across the river
Environment
Encourage the creation of the Municipal Ecological Park of Maxixe, a project proposal made by the Universidade Pedagógica—Maxixe
Constant monitoring of the environment guarantees a prompt reaction to new arising issues
Protection of flora and fauna in the area surrounding the research center
Monitoring of the bay ecosystem and protection from excessive fishery and use of illegal methods (e.g. explosive)
Introduction of sustainable source of energy and of an electricity system that will lower the increasing use of wood as the main source of energy
Economic
Introduction of a new economic activity in the area
Yearly average profit coming from rent and other services offered during the touristic season is estimated to be around €90,000
The initial investment is expected to be recovered in less than 5 years and 6 months
During the calculation of the investment Net Present Value, the use of a discount rate smaller than 14.69% resulted in a positive outcome of the analysis
The underlying hypothesis of being inside a natural reserve guarantees a donation equal to €20,000 to support local communities
Exploitation of every period of the year, maximising the use of the receptive structure

Authors: L. Baiardi, M. Talliani

Acknowledgements The Mo.N.G.U.E. project is funded by the Polisocial Program and is carried out jointly with the 'Pedagógica Universidade de Moçambique, Delegação de Maxixe-UniSaF and The Sagrada Familia Congregation. The following are supporters with a declared interest: Pedagógica the Universidade de Moçambique, the town of Maxixe, the Diocese of Inhambane, the Italian Embassy in Maputo.

References

IFAD (International Fund for Agricultural Development) (2014) Investing in rural people in Mozambique. https://www.ifad.org/documents/10180/bf1817c4-7061-40d6-9291-4512691f15fd

IHS (Institute for Housing and Urban Development Studies) (2017) Urbanization in Mozambique—assessing actors, processes, and impacts of urban growth. http://www.citiesalliance.org/sites/citiesalliance.org/files/Urbanization%20in%20Mozambique.pdf

Instituto Nacional de Estatística (2015) Anuário Estatístico 2015 - Moçambique. Instituto Nacional de Estatística, Maputo, Mozambique

Monga C, Lin JY (2015) The Oxford handbook of Africa and economics: Volume 1: context and concepts. Oxford University Press, New York, USA

Monti A, Romano MG (2010) Studi di fattibilità di progetti complessi. In: Mangiarotti A, Tronconi O (eds) Il progetto di fattibilità. Analisi tecnico-economica e sistemi costruttivi. McGraw-Hill, Milano (IT), pp 39–50

Ollivier T, Giraud PN (2011) Assessing sustainability, a comprehensive wealth accounting prospect: an application to Mozambique. Ecol Econ 70(3):503–512

PARPA (2011) International Monetary Fund "Republic of Mozambique: poverty reduction strategy paper", June 2011, IMF country report no. 11/132; https://www.imf.org/external/pubs/ft/scr/2011/cr11132.pdf

Porter A, Bohl D, Kwasi S, Donnenfeld Z, Cilliers J (2018) Prospects and challenges: Mozambique's growth and human development outlook to 2040. http://dx.doi.org/10.2139/ssrn.3099373

Scott J (2015) The risks of rapid urbanization in developing countries. https://www.zurich.com/en

Thompson DA (2013) Constructing a history of independent Mozambique, 1974–1982: a study in photography. Kronos 39(1):158–184

UN-Habitat (United Nations Human Settlements Programme) (2015) Towards an African urban agenda. Nairobi, Kenya

UNICEF Mozambique, Water, sanitation & hygiene. http://www.unicef.org.mz/en/our-work/what-we-do/water-sanitation-hygiene/

WHO (World Health Organization) (2014) Mozambique. Urban health profile. http://www.who.int/kobe_centre/measuring/urbanheart/mozambique.pdf

World Peace Foundation (2015) Mozambique: civil war. https://sites.tufts.edu/atrocityendings/2015/08/07/mozambique-civil-war/

Past, Present and Future

Introduction

Cinzia Talamo, Niccolò Aste, Corinna Rossi, Rajendra Singh Adhikari

Archaeology is the study of the material remains of the human past (artifacts, technology, buildings and structures), and can offer important information on how humankind interacted with the environment.

Archaeological remains suffer from three types of destructions: the first is due to the passing of time, during which decayed traces of the past might or might not be absorbed into progressively new installations; the second is due to voluntary damages, exacerbated during wars and conflicts; the third is due to archaeological excavations.

The first type of destruction depends partly on the natural decay and partly on misuse and neglect. The second is unfortunately well-known: recent conflicts that saw the cultural heritage blatantly targeted to hit specific populations, their possessions and their cultural identity are just the latest examples of a sadly long-running practice. Properly documenting endangered archaeological sites is currently perceived as one of the priorities in conflict zones, in order to preserve the historical and cultural identity of the local populations.

The third type of destruction is not always taken into account: archaeological excavation is destructive by definition, as archaeologists physically remove layers of remains and remove items from their finding spot. Part of the meaning and function of the finds is actually provided by their context: failing to properly document the original connection between them heavily hampers our knowledge, as it reduces our chances to reconstruct the history of objects and sites.

Modern technology offers elaborated and efficient instruments able to perform extremely accurate surveys of these remains, ranging from large buildings to small items and from inorganic to inorganic finds, to the point that an entirely new field of research developed, that of archaeometry, able to reveal aspects and characteristics of the material culture that is being investigated that have ben so far invisible or undectable, both to the specialists and to the greater public.

This section presents some research projects related to analysis and low cost web based digital tools for archaeology developed to be applied in North Africa, and concludes the volume with a contribution focussing on the survival and transmission of models throughout space, time and cultures.

Low-Cost Digital Tools for Archaeology

Luca Perfetti, Francesco Fassi and Corinna Rossi

Abstract Modern technology offers elaborated and efficient instruments capable of performing extremely accurate surveys of architectural and archaeological remains. However, not all of them can be used everywhere: archaeological missions might be constrained by logistics, environmental and, especially, financial restrictions. This issue is especially felt by archaeological missions currently operating in the Middle East and Africa. The research team of the ERC project LIFE (CoGrant 681673) has been successfully experimenting with the use of low-cost instruments to achieve equally accurate results.

Keywords Photogrammetry · Low-cost 3D · Fisheye · Action camera

1 Introduction

Modern 3D scanner technology is widely used throughout archaeology. Accessing the latest technological tools is often directly related to the availability of substantial funding and to the actual possibility of being able to use them on the field. For these reasons, operators not endowed with sufficient economic resources and/or who are engaged in difficult logistic and environmental situations run the risk of encountering a number of difficulties.

However, nowadays, a new generation of 'image-based survey techniques' can be used to achieve equally satisfactory results in terms of accuracy and completeness of the results: those relying on low-cost instruments and procedures, based on the concepts of flexibility, portability and ease of use. For archaeological projects and for complex conditions of acquisition, as well as, in general, for the vast field of cultural heritage, this type of technique is rapidly becoming the main solution for documenting excavations, findings and archaeological remains.

L. Perfetti (✉) · F. Fassi · C. Rossi
Architecture, Built Environment and Construction Engineering—ABC Department, Politecnico di Milano, Milan, Italy
e-mail: luca.perfetti@polimi.it

N. Aste et al. (eds.), *Innovative Models for Sustainable Development in Emerging African Countries*, Research for Development, https://doi.org/10.1007/978-3-030-33323-2_12

2 Photogrammetry: The Low-Cost Digital Solution

In the last century, photogrammetry was mainly used to survey the territory and produce cartography. Following the first decades of development of photogrammetry, mainly in the aerial field, close-range photogrammetry for archaeology or architecture could not fully assert itself, since similar technologies and the manual procedures of orientation and stereoplotting were perceived as time-consuming, and consequently very expensive.

Seemingly, the richness of detail and the complexity of artistic free-forms require a high number of high-resolution images and complex 3D modelling procedures in order to be able to adequately describe the shape of the object. For these reasons, until recently, the use of manual or semi-manual photogrammetric methods have prevented or severely limited complete three-dimensional surveys. This was obviously true in the period of analogue photogrammetry (before 2000), but remained so with the advent of digital photography (2000–2010), since photogrammetric processing operations, in one way or the other, remained a manual operation (Fassi and Campanella 2017).

In recent years (from 2010 onwards), there have been important developments in close-range photogrammetry thanks to the application of Computer Vision algorithms, able to automatize the processes of tie-point identification, camera calibration, image orientation and dense DSM reconstruction (image matching), thus allowing for accurate 3D reconstructions.

These methods are called 'image-based techniques': they integrate photogrammetric concepts of image orientation and camera calibration with CV algorithms for key point identification and dense image matching (Fassi et al. 2013). Today, both in literature and in practice, these methods are also called 'image scanning techniques', or 'photoscanning'. We find this terminology misleading, but the nomenclature underlines the main important tasks that they propose: the 3D dense reconstruction of the object as a 3D point cloud, comparable to laser scanner data in terms of achievable resolution and accuracy in surface description.

Furthermore, image-based methods have additional positive aspects:

- *Flexibility*: with the same camera, it is possible to survey both large objects (such as a building or entire archaeological sites) and small objects. The point cloud density depends solely on the image's resolution and the capture geometry.
- *Accuracy knowledge* point by point. The 3D position of each reconstructed point is a mathematical operation; for this reason, it is possible to know the accuracy of each point.
- *Metric chromatic mapping* in the 3D model.
- *Self-calibration of the camera system* which allows the use of normal photographic systems, as well as alternative image systems such as fisheye lenses, panorama cameras, multi-camera systems, low-cost action cameras or even micro cameras (including mobile phone cameras).

In this way, the photographic acquisition is freed from the photogrammetric constraints of the past: the great flexibility and 'easy acquisition', typical of photography allow for freedom of movement around the object and a multiscale acquisition while also offering the possibility to survey both very complex and large object as well as tiny decorations.

Another important advantage is that all these characteristics also allow for low-cost surveys in cases of emergency, where one can use common tools and obtain a metric result nonetheless. The final accuracy of the process can always be calculated (and this is the photogrammetric meaning of the word 'metric'!); it will be numerically similar to the resolution of the images with some pejorative variations that are strictly dependent on photographic quality.

3 Case Studies

The 3D survey group, as part of the ERC Project LIFE, is currently testing low-cost digital tools on two archaeological sites, characterized by different environmental and logistic conditions.

Umm al-Dabadib is located on the outskirts of the Kharga Oasis, one of the five largest oases in Egypt's Western Desert. It consists of a vast and well-preserved Roman installation including an inhabited area served by an extensive agricultural system, certainly active in the Third and Fourth Centuries AD. The site was probably abandoned at the beginning of the Fifth Century and, apart from a short-lived occupation dating back to the early Twentieth Century, it survived up to today relatively intact and virtually unknown.

Briefly described in the early Twentieth Century (Beadnell 1909), its real extent was only revealed in 1998 (Rossi 2000); it was then studied for the first time in some detail between 2001 and 2007 by the North Kharga Oasis Survey (NKOS, Rossi and Ikram 2018). Between 2012 and 2014, its agricultural system was the object of a specific study by the Old Agricultural Sites and Irrigation Systems (OASIS) project of the MUSA Centre of the Federico II University of Naples (Fassi et al. 2015; Rossi 2016; Rossi forthcoming-a); it is now the object of the multi-disciplinary project Living In a Fringe Environment (LIFE), jointly carried out by the Politecnico di Milano and the MUSA Centre and funded by the ERC CoGrant 681673 (Rossi 2017; Rossi forthcoming-b; Rossi and Fiorillo 2018; Rossi and Magli forthcoming).

This remote site is located at a distance of over 50 km from the nearest inhabited centre, and therefore, any activity must be planned in the absence of water, electricity and Internet. The strong winds that periodically batter the area heavily affect the integrity of any electronic device that is carried on site. Drones are not an option, as they are prohibited in Egypt; moreover, Umm al-Dabadib lies within a military area, where further restrictions apply.

In this case, the survey of the entire Fortified Settlement covering a densely built-up surface of ca. 100 × 100 m, and reaching peaks of over 10 m with the central Fort, was carried out entirely by photogrammetry (Fassi et al. 2015).

The survey work was divided into two phases—the survey of the exteriors and the survey of the interiors. To complete the survey of the external part, over 5100 photographs were necessary. The area was divided into sub-areas corresponding to different housing units. The target was to survey the remains paying special attention to the architectural structures, with the future aim of understanding the original dimension and form of the settlement. For this goal, it was necessary to survey not only all the external parts but also the part just below the surface such as vaults, niches and holes.

The survey of open and close spaces together, thus required a multiscale approach, implemented thanks to a Canon 5D Mark III with a fixed 35 mm F2 lens and a Canon G1X. The professional reflex camera was mainly used for the main buildings and the largest areas, due to the high resolution (22.3 Mp), the high-quality lenses and the major configuration possibilities. The compact one was used for the smaller buildings, for narrow spaces and when it was necessary to climb over some structures.

All the survey block was elaborated separately using Agisoft Photoscan in order to produce a high-resolution point cloud capable of representing in 3D the entire settlement. All the different sub-areas were registered together using markers, with the foresight to detect parts of neighbouring buildings during the survey of each subarea. The use of markers can be avoided in case of an emergency, but it is mandatory to compute the correct camera auto-calibration, to scale the final model correctly and to check the final metric results (Fassi et al. 2015).

The final results of this survey consisted of a complete and very dense point cloud of the settlement, consisting of about 650 million points, with a spatial resolution of circa 5 mm. It is a kind of very high-density LIDAR point cloud, significantly higher than the classical resolution suitable for architectural/archaeological representation which is typical in these cases.

The second phase was to survey the internal spaces of the castle, consisting of little rooms and passages entirely or partially buried under the sand. For this reason, these spaces are dark, difficult to reach from the outside and impose strong limitations to the operator moving in them.

The survey of narrow spaces is an important topic nowadays, and many approaches are being tested in different environments in order to check the capability of 3D reconstruction, reliability of the methods and quality of the final results (Covas et al. 2015; Barazzetti et al. 2017a, b; Mandelli et al. 2017; Pepe et al. 2018). One of the proposed methods is fisheye photogrammetry, and right here, (2014) this was tested for the first time by the research group. The wide angle of the lenses allows users to survey this type of spaces based on a lower number of photos (Perfetti et al. 2017).

To overcome the lighting problem, we used speedlight when the spaces were small enough to avoid the problem of image vignetting. In this case, a light painting technique was used to uniformly illuminate the scenes. Fisheye photogrammetry was used to survey the staircase of the tower and the little rooms at the ground level of the castle. The use of the consumer camera Canon EOS350D coupled with a fixed 8 mm focal length and no ideal illumination devices was enough to get the geometry of the spaces with an accuracy suitable for a 1:50 scale representation even if the quality of the texture is insufficient for a qualitative description of the spaces. In order to register external and internal spaces, it was necessary to have a good overlapping zone between the two areas; it was a difficult task due to the narrow passages that

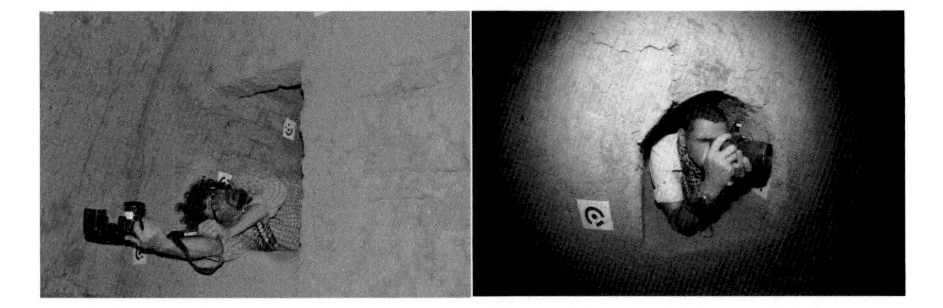

Fig. 1 Survey of the interiors of the castle in Umm el Dabadib. Fisheye photogrammetry was necessary to model the narrow and dark spaces, not easily accessible because of their reduced dimensions, as well as because sand and debris from the upper parts of the building, that collapsed in the past, flowed in and occupied accesses and corridors. Francesco Fassi and Alessandro Mandelli on the field in 2014 and 2015, surveying the windowless ground level of the castle. The same method was used to survey the building's staircase, which served all five floors, and revolved around a central pillar

connect the two areas. The use of reference markers was also in this case mandatory in order to guarantee the success of the operation (Fig. 1).

In the survey of the Fortified Settlement of Umm al-Dabadib, using a laser scanner was impossible due to the extremely complex logistic and environmental constrains (the electricity supply on site is guaranteed only by a generator, and the frequent sand-laden wind would damage the fragile components of expensive machines).

Even if using a scanner had been possible, the layout of the spaces, the wealth of details and the presence of narrow spaces would have made that type of survey longer and more difficult, as the acquisition would have implied a high number of scans, the need to continuously re-position the laser, together with all the ensuing registration problems.

For all these reasons, photogrammetry was the winning choice for this type of application, allowing complete and high-quality results in the open spaces. The 'almost impossible' challenge to perform the 3D survey of narrow spaces became possible thanks to the use of low-cost consumer sensors.

The double-phased survey was necessary to complete the 3D reconstruction of a very complex environment allowing for a full comprehension of the spaces (Fig. 2). The ensuing results allowed the team to start a metrological research into the planning and building methods used by the ancient builders, which otherwise would have been impossible (Rossi and Fiorillo 2018).

Another application of digital imaging for archaeology has been tested on the field at Saqqara (Egypt), on the joint Dutch-Italian excavation of the New Kingdom necropolis conducted by the Museo Egizio, Turin and the Rijksmuseum van Oudheden, Leiden. The area host a number of tombs of high officials who served the pharaohs of the Eighteenth and Nineteenth Dynasties, surrounded by minor burials and covered by later layers of occupation; they were progressively discovered and excavated from the '80s onwards (Martin 1989, 1991, 1997, 2001; Raven 1991, 2005;

Fig. 2 A 'radiography' of the castle showing the relationship between internal and external spaces of the castle (elaboration by Fiorillo 2018)

Raven and van Walsem 2014). The area currently under excavation stretches north of the Eighteenth Dynasty Tomb of Maya and appears to include a slightly later middle-sized tomb and several Ramesside burials that reused older underground structures.

The proximity to the Nile Valley and the presence, at a short distance, of a fully equipped dig house allows the team to benefit from regular access to electricity, as well as Internet connection. Other environmental and legal restrictions apply here as well: drones are banned, and the sandy wind tends to damage fragile equipment. Photogrammetry was also chosen in this case as the main working tool.

Specific to this project is the need to record in 3D every single context that is excavated, to construct a virtual digging diary of the activities, to be later examined, to visualize the layers that have been removed and double-check meaningful details. Surveying contexts means surveying the same area several times in a row as it evolves and measuring relatively small flat surfaces combined with structures and small remains. In this case, therefore, the main goals are as follows:

- *Global georeferencing all the surveys together*, in order to visualize all the excavation phases in the correct position. This operation must be very accurate because the positioning of the different contexts in the *Z*-axis is crucial. Most of the time, the contexts consist of very thin layers, and millimetric accuracy is necessary in order to place them in their correct position and avoid false intersections or misalignments.
- *Very high-resolution survey*, in order to distinguish different contexts, a very large representation scale is necessary. The different layers can be distinguished only by different granulometry of the materials or small colour differences that should

be shown as to allow for the future visualization, study and interpretation of the data.

- *Near real-time elaboration*, in order to facilitate the digital cataloguing of all the information, the survey team should produce the 3D models of the context quickly, ideally in parallel to the excavation.

Close-range photogrammetry is the only possible method that allows the team to satisfy all of these requirements, both individually and as a group. The most important aspects of this method, in fact, correspond closely to the necessities listed above. Its main characteristics are as follows:

- great flexibility in terms of achievable resolution, thus able to cover all the various needs of an archaeological excavation, ranging from relatively vast areas, to groups of finds (cf. Figs. 3, 4 and 6);
- the possibility of placing thin and tiny contexts in their correct absolute and relative position (Figs. 4 and 5);
- the possibility of using exactly the same instruments to perform the survey of objects of significantly different shapes and dimensions (cf. Fig. 6, showing the extremely detailed survey of a nearly vanished graffito on a limestone block which was crucial for the interpretation of the later use of that tomb);
- the possibility to achieve very large representation scales, up to 1:1 (see Figs. 4 and 6);

Fig. 3 Alessandro Mandelli and Luca Perfetti using a photographic crane to survey the area just excavated by the Dutch-Italian Mission to Saqqara, April 2019

Fig. 4 The textured model of a context. It can be visualized up to a representation scale of 1:1 thanks to a texture resolution of 0.5 mm. This represents a useful tool for archaeologists, who can virtually revisit the site and double-check details weeks or months after the excavation, in a very realistic way

- being a fast procedure, the possibility of repeating the survey many times throughout the same day, following the excavation phases step by step. The same find, for instance, can be surveyed several times as it slowly emerges from the surroundings, thus recording with precision its original context. This is extremely important, since an archaeological excavation is a destructive process, that physically removes what is found and separates the finds from one another and from their original context. What is not recorded on the spot is lost, both in terms of information and in terms of precision.

The described approach is what we use during the survey activity to support the excavation in Saqqara. A topographic network around the excavation area was measured to geo-reference all the micro-surveys together—a Canon 5D Mark III with 35 mm or 20 mm lenses was the camera used for photographic acquisition. For the smaller contexts, the resolution of 0.1 mm was ensured. A minimum of four targets were positioned around the scene using four long nails to anchor them into the sand for the short survey time (cf. Fig. 4). The markers were measured with a total station (Topcon ES62) and removed after every single survey operation.

For larger areas, the camera was mounted on a transportable photographic crane reaching a height of 3 m to capture the scene from a high vertical point of view with a GSD of less than 1 mm (Fig. 3). It was necessary to adopt this system due to the impossibility of using UAVs.

Fig. 5 A temporal sequence of excavated contexts from the 2018 excavation of the Dutch-Italian archaeological mission to Saqqara, of the Museo Egizio, Turin and the Rijksmuseum van Oudheden, Leiden: a number of 'embalming caches' are found and recorded around a burial shaft. Embalming caches contained the remains of the materials used during the mummification process and were buried near the tomb

Fig. 6 A very high-resolution model of a graffito from the tomb of Horemheb at Saqqara. The original images were taken with 0.03 mm. The matched model is in full resolution and allows us to read 3D details hardly visible to the naked eye

4 Conclusions

The presented case studies aim to demonstrate how nowadays it is possible to survey complex environments using low-cost tools and yet fully exploit the potential of image-based techniques. Moreover, not only can photogrammetry replace other survey methods (e.g. laser scanning) and give similar results, but in some cases, such as those described here, photogrammetry actually becomes the best and only viable solution.

The possibility of using the same camera to survey both architecture and small objects and offer the adequate resolution for each case, the flexibility in their use, and their low-cost characteristics represent the key factors for preferring this method to the classical ones of a range-based 3D survey. Different rules and procedures must be carefully followed in order to obtain metric and accurate results. Topographic support is highly recommended to geo-reference all the data together, to scale the different surveys, and to assess the final result. The use of markers or 'natural constraint points' is mandatory for the global correctness of the photogrammetric process (Fassi and Campanella 2017).

References

Barazzetti L, Previtali M, Roncoroni F (2017a) Fisheye lenses for 3d modeling: evaluations and considerations. In: International archives of the photogrammetry, remote sensing and spatial information sciences, Nafplio, Greece, vol XLII-2/W3, pp 79–84. https://doi.org/10.5194/isprs-archives-xlii-2-W3-79-2017

Barazzetti L, Previtali M, Roncoroni F (2017b) 3D modelling with the Samsung gear 360. In: International archives of the photogrammetry, remote sensing and spatial information sciences, Nafplio, Greece, vol XLII-2/W3, pp 85–90. https://doi.org/10.5194/isprs-archives-xlii-2-w3-85-2017

Beadnell HJL (1909) An Egyptian oasis, London

Covas J, Ferreira V, Mateus L (2015) 3D reconstruction with fisheye images strategies to survey complex heritage buildings. In: 2015 digital heritage, Granada, pp 123–126. https://doi.org/10.1109/digitalheritage.2015.7413850

Fassi F, Campanella C (2017) From daguerreotypes to digital automatic photogrammetry. Applications and limits for the built heritage project. In: International archives of the photogrammetry, remote sensing and spatial information sciences, vol XLII-5/W1, pp 313–319. https://doi.org/10.5194/isprs-archives-XLII-5-W1-313-2017

Fassi F, Fregonese L, Ackermann S, De Troia V (2013) Comparison between laser scanning and automated 3d modelling techniques to reconstruct complex and extensive cultural heritage areas. In: International archives of the photogrammetry, remote sensing and spatial information sciences, vol XL-5/W1, pp 73–80

Fassi F, Rossi C, Mandelli A (2015) Emergency survey of endangered or logistically complex archaeological sites. In: International archives of the photogrammetry, remote sensing and spatial information sciences, vol 40, no 5, pp 85–91

Mandelli A, Fassi F, Perfetti L, Polari C (2017) Testing different survey techniques to model architectonic narrow spaces. In: International archives of the photogrammetry, remote sensing and spatial information sciences, XLII-2/W5, pp 505–511. https://doi.org/10.5194/isprs-archives-XLII-2-W5-505-2017

Martin GT (1989) The Memphite tomb of Ḥoremḥeb, commander-in-chief of Tut'ankhamūn, I: the reliefs, inscriptions, and commentary. Egypt Exploration Society, Excavation Memoir 55, London

Martin GT (1991) The hidden tombs of Memphis, London

Martin GT (1997) The tomb of Tia and Tia, a royal monument of the Ramesside period in the Memphite Necropolis, London

Martin GT (2001) The tombs of three Memphite officials: Ramose, Khay and Pabes, London

Pepe M, Ackermann S, Fregonese L, Fassi F, Adami A (2018) Applications of action cam sensors in the archaeological yard. In: International archives of the photogrammetry, remote sensing and spatial information sciences, vol XLII-2, pp 861–867. https://doi.org/10.5194/isprs-archives-XLII-2-861-2018

Perfetti L, Polari C, Fassi F (2017) Fisheye photogrammetry: tests and methodologies for the survey of narrow spaces. In: International archives of the photogrammetry, remote sensing and spatial information sciences, vol XLII-2/W3, pp 573–580. https://doi.org/10.5194/isprs-archives-XLII-2-W3-573-2017

Raven MJ (1991) The tomb of Iuredef. A Memphite official in the Reign of Ramesses II, London and Leiden

Raven MJ (2005) The tomb of Pay and Raia at Saqqara, London and Leiden

Raven MJ, van Walsem R (2014) The tomb of Meryneith at Saqqara. PALMA 10, Brepols

Rossi C (2000) Umm el-Dabadib, Roman settlement in the Kharga Oasis: description of the visible remains. With a note on 'Ayn Amur. Mitteilungen des Deutschen Archäologischen Instituts Kairo 56:235–252

Rossi C (2016) Italian mission to Umm al-Dabadib (Kharga Oasis): season 2014—preliminary report. Mitteilungen des Deutschen Archäologischen Instituts Kairo 72:153–172

Rossi C (2017) Survey, conservation and restoration in Egypt's Western Desert: combining expectations and context. Restauro Archeologico 2:4–19
Rossi C (forthcoming-a) Italian mission to Umm al-Dabadib (Kharga Oasis): season 2015—preliminary report. Mitteilungen des Deutschen Archäologischen Instituts Kairo 74
Rossi C (forthcoming-b) Searching for the right words: what happened in Kharga in the IV century AD? In: Proceedings of the colloquium Marges et frontières occidentales de l'Égypte, Cairo, 2–3 Dec 2017, Bibliothèque d'Étude
Rossi C, Fiorillo F (2018) A metrological study of the Late Roman Fort of Umm al-Dabadib, Kharga Oasis (Egypt). Nexus Netw J 20(2):373–391
Rossi C, Ikram S (2018) North Kharga Oasis survey. Explorations in Egypt's Western Desert. British Museum Publications on Egypt and Sudan 5, Leuven
Rossi C, Magli G (forthcoming) Wind, sand and water: the orientation of the Late Roman Forts in the Kharga Oasis (Egyptian Western Desert). In: Magli G, Antonello E, Belmonte JA, César González-García A (eds) Archaeoastronomy in the Roman world, Berlin

Digital Workflow to Support Archaeological Excavation: From the 3D Survey to the Websharing of Data

Corinna Rossi, Cristiana Achille, Francesco Fassi, Francesca Lori, Fabrizio Rechichi and Fausta Fiorillo

Abstract Archaeology has recently seen a rise in the use of digital tools and new technologies. However, in many cases, innovative tools are used to perform old operations, and their potential is not fully exploited to achieve equally innovative results. Moreover, in the practice of archaeological excavations, the collection of digital data proceeds alongside the collection of classic paper archives, thus prompting the necessity to find a way to combine different sets of data. The research team engaged in the ERC project LIFE (CoGrant 681673) is working on the identification of the most effective survey methods in relation to specific logistic and environmental conditions, as well as on the possibility to efficiently combine digital and paper archives, and is testing the results on two archaeological excavations in Egypt.

Keywords WEB-BIM · BIM3DSG · 3D models · Reality-based modelling · Sharing · Big data · Informative system · Archaeology

1 Introduction

In the last ten years, new technologies have been gaining momentum in archaeology; however, their use and distribution are still rather uneven. Both the acquisition (thanks to the various available options such as laser scanners, UAV and photogrammetry) and the classification of data in an electronic format are becoming increasingly commonplace, but the interaction between the various sets of data is often patchy. This is due to a combination of issues, including the understandable need to avoid substantial disruptions to the consolidated archaeological workflow, as well as a number of technical difficulties relating to the interaction between digital and paper archives. The use of pen and paper on an archaeological excavation can be supported but cannot be entirely replaced by digital means; the issue here is not the presence of different sets of data, but the fact that these sets of data do not communicate in an effective or efficient way. Therefore, the challenge is to identify the most efficient

C. Rossi (✉) · C. Achille · F. Fassi · F. Lori · F. Rechichi · F. Fiorillo
Architecture, Built Environment and Construction Engineering—ABC Department, Politecnico di Milano, Milan, Italy
e-mail: corinna.rossi@polimi.it

N. Aste et al. (eds.), *Innovative Models for Sustainable Development in Emerging African Countries*, Research for Development, https://doi.org/10.1007/978-3-030-33323-2_13

survey methods both in terms of implementation on the field and of post-processing and eventually set up a comprehensive informative system which would be able to combine different types of data.

2 Fieldwork and Digital Tools

On the field, pencil and paper still play a fundamental role and will continue to do so. The vast majority of archaeological missions still produce a large amount of data on paper, from which, at a later stage, all the relevant information will be collected in order to reconstruct the stratigraphy and the evolution of the site. The study of specific finds, sometimes, takes place years after their retrieval; in general, the post-fieldwork processing can take a long time, and sometimes never reaches an end, thus producing a variable amount of so-called grey literature, that is, data that is collected but never published (Fig. 1).

Digital tools are gaining space in the archaeological practice, but they are often confined to specific areas, their potential not fully exploited. For instance, 3D models

Fig. 1 Data may be acquired in a digital manner; however, on the field, pen and paper represent the most useful and convenient tools to take notes and check cross-references. In Egypt, sand, wind and strong sunlight make the use of tablets on the field rather uncomfortable. The issue is not necessarily whether or not the technology is available, but whether or not it is really useful in that specific circumstance. Archaologists comparing written notes with one another, during the 2018 Dutch-Italian excavation at Saqqara of the Museo Egizio, Turin and the Rijksmuseum van Oudheden, Leiden

of items are used just as 'better' (more attractive and efficient) images in comparison with photographs, but they are rarely exploited to really enhance specific characteristics of the object which would otherwise be invisible or undetectable (e.g. Rossi and Fiorillo 2018).

One aspect that is gaining importance is the use of photogrammetry to obtain orthoimages of excavated areas (e.g. Verhoeven et al. 2012; Fregonese et al. 2016). Seemingly, electronic databases are now widespread; even if they all respond to some basic requirements, they are generally run independently by each archaeological mission and tailored to their specific needs. Another issue is represented by the fact that some obviously expected results imply rather complex technical solutions: for instance, achieving the connection between items and space (so easily noted on paper) means merging 3D, GIS and databases into an informative system (a complex operation).

The overall impression is that there is ample room to find a more productive and proactive role of digital imaging within archaeological practice, but some basic conditions must be respected.

Archaeological excavations rely on long-established practices, drawing from a common corpus of rules and conventions, interspersed with specific variations depending on the habits of the specialists involved in the excavation and the post-excavation phases. Especially in the case of long-running projects, the introduction of new methods and tools is certainly welcome, but can be problematic if they disrupt the consolidated workflow.

Any major change would be a matter of modifying not only personal habits, but also the methodologies with which data is collected, classified and studied thus far. Therefore, the introduction of any new tool must be carefully evaluated in terms of impact on consolidated practice, as to avoid that the costs of its adoption surpass the benefits.

3 Establishing Connections

The most interesting and productive aspect of a closer interaction between archaeology and digital imaging is the possibility to virtually construct or reconstruct lost contexts and severed ties.

An archaeological excavation is a destructive activity: progressively, archaeologists physically remove the layers that accumulated over the centuries and separate forever the items that are found during the excavation from their original context. When the components of the stratigraphy are divided, some items start a new life, becoming objects to be studied and possibly put on display (Fig. 2).

This process of destruction and separation is irreversible. The only surviving link to the original context is the information recorded during the excavation phase; when this is not available (either because an object was found during illicit digging activities, or because the excavation reports remain unpublished), the process of

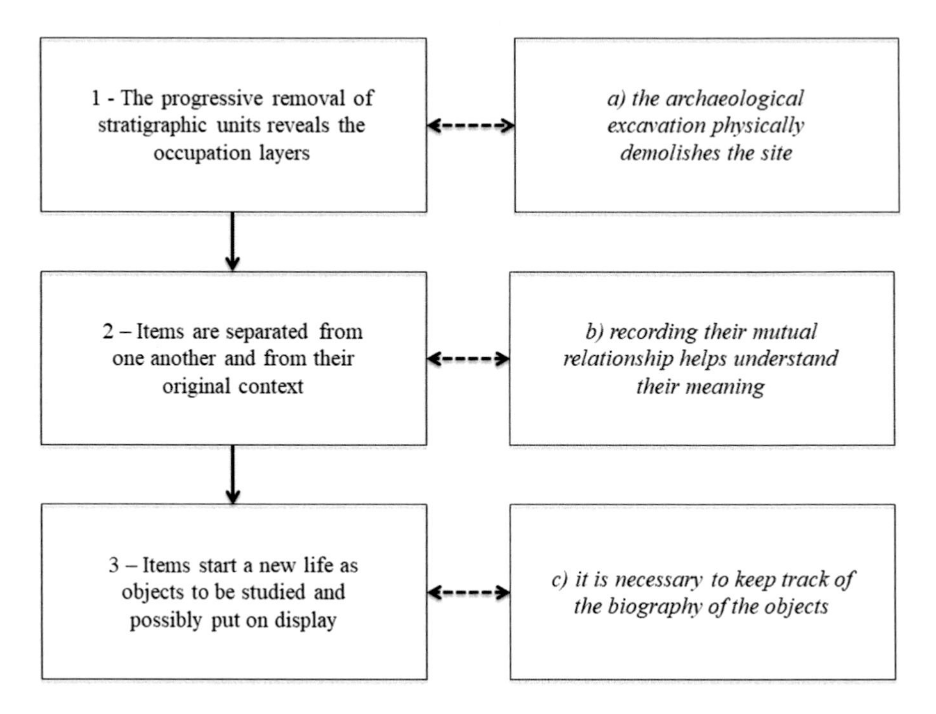

Fig. 2 Archaeology as a destructive process: actions, consequences and countermeasures

detachment of the object from its original setting, and therefore its original identity, reaches its maximum extent (Del Vesco 2018).

Digital imaging, in itself, represents a way to document in an efficient and effective way a number of aspects of both the excavation process and the items that are retrieved. Potentially, it may provide an efficient basis on which to create connections among the various types of data, in order to reconstruct the lost context and re-trace the complete biography of the objects (Betrò 2011; Greco 2018).

Digital records offer the additional benefit of being able to share information easily; this means reducing the danger of generating grey literature, as well as offering the possibility to create or re-create virtual connections that would be impossible to achieve in the real world.

In conclusion, connectivity is the keyword around which to construct a successful dialogue between archaeological practice and new technologies, aiming at responding to the needs of archaeologists and, at the same time, exploiting the potential of digital imaging and records.

4 Surveying in 3D an Archaeological Excavation

Archaeological fieldwork is traditionally recorded on context sheets, one for each context retrieved during the excavation process, which is filled by hand with all the relevant information (identification code, retrieval date, location, dimensions, material, etc.). The relative position of each context in relation to the adjacent contexts is carefully noted; all of this information is later conflated into the so-called Harris Matrix, a scheme that summarizes the mutual relationships of all contexts that have been retrieved.

Although, in a way, this passage represents the three-dimensionalization of the excavation data, it leads to the construction of a vertical section of the stratigraphy, a two-dimensional representation. The traditional archaeological method, therefore, skirts the third dimensions, but never really deals with it as it moves mainly on paper. The possibility to produce 3D surveys and models of the excavated remains, thus, offers fresh possibilities of investigation and practical applications that are likely to produce interesting results in near future.

While the 3D survey of standing remains offers the same advantages that are already known from the more general field of digital techniques applied to the cultural heritage, the potential of surveying in 3D the stratigraphy of an excavation is still being explored. The research team of the ERC project LIFE is working on this subject in collaboration with the Museo Egizio, Turin, at the excavation of the New Kingdom tombs of Saqqara (Egypt), led by Rijksmuseum van Oudheden, Leiden and the Museo Egizio, Turin.

The concession of the Dutch-Italian mission includes a number of large tombs built for themselves by high-ranking officials who lived in the period during the Eighteenth and the Nineteenth Dynasties of the New Kingdom (Martin 1991). Among them, there was Maya, Treasurer of Tutankhamun, and Horemheb, who built a tomb for himself there when he was a powerful general of the Egyptian army; he later became pharaoh and was eventually buried in the Valley of the Kings, on Luxor's West Bank.

The upper levels of the area currently under excavation, located to the north of the Tomb of Maya, contain evidence of later occupation, including some small Rameside chapels built alongside re-used funerary shafts, and the later, feeble remains of domestic occupation dating back to the Coptic Period (Del Vesco et al. 2019). All these remains are being surveyed in 3D by photogrammetry, thus offering archaeologists the chance to construct something of a virtual digging diary, to which they can go back to check the appearance of the excavation in any given day. The use of photogrammetry means that the ensuing 3D models have a realistic appearance which, if matched with a very high resolution, allows for the visualization of a wealth of small details, some of which might have been overlooked on the spot. This is proving an extremely useful tool during the post-fieldwork processing of the data, which takes place, by definition, after the fieldwork and away from the site.

5 Handling the Data

Once the 3D data has been collected and processed, the problem is how to let it interact with other types of data (texts, images, etc.) containing other information. This is a more general problem, if course, that is not specific to the archaeological realm.

Starting in 2010, the 3D survey group of the Politecnico di Milano developed a prototype of an HBIM (Building Information Model for Cultural Heritage) system for the maintenance of the Veneranda Fabbrica del Duomo di Milano (Fassi et al. 2011). This experiment worked as a pilot project to develop a more general information system (BIM3DSG), to be proposed as a standard fruition and valorization procedure in the world of cultural heritage, using 3D as the basis of informative systems (Rechichi et al. 2016). The expected result was to provide a tool that could enhance the potential of using a virtual digital model in the cultural heritage sector, particularly for the restoration, extraordinary and ordinary maintenance of a historical and artistic monumental complex. Later, the system was also applied to the Basilica di San Marco in Venice (Fassi et al. 2017; Adami et al. 2018), for the conservation practices of Pietà Rondanini (Mandelli et al. 2017) and over the last few months for the conservation of large architectonic environmental UNESCO heritage sites such as the Sacri Monti of Piedmont and Lombardy (Tommasi et al. 2019).

BIM3DSG is created for the advanced management and 3D visualization of heterogeneous models characterized by a high geometrical complexity, as is common in the field of cultural heritage (Fassi et al. 2014). The system is divided in two parts. The first is conceived to be mainly used by professionals and 3D specialists, and it is developed into the modelling software and aims to add or modify 3D models (point cloud, nurbs and mesh with or without texture). The second part is conceived for all other users and allows for the use of the system via the web. It requires only a web browser and is specifically designed to also be used on mobile devices such as laptops, tablets and smartphones, even those characterized by low hardware resources. Both sections allow the user to access the interesting parts, zones, sectors, areas, in addition to the whole model; the selection of desired objects can be achieved through a variety of search functions or can be obtained automatically through spatial relationships (Fassi and Parri 2012).

A sample of possible operations that can be carried out within the system are (i) to manually compute distance measurements and automatically measure surface area, volume and coordinates of every object; (ii) to add/edit/view user information; (iii) to attach external files, such as photos, videos, documents and dwg files associated with one or more objects or models and (iv) to add/edit/view maintenance, restoration and building site activities with all related information. All these operations can be carried out via a web browser.

The core of the system is a dynamic database that contains all the data and automatically manages the use of the system via web, both in reading and in writing

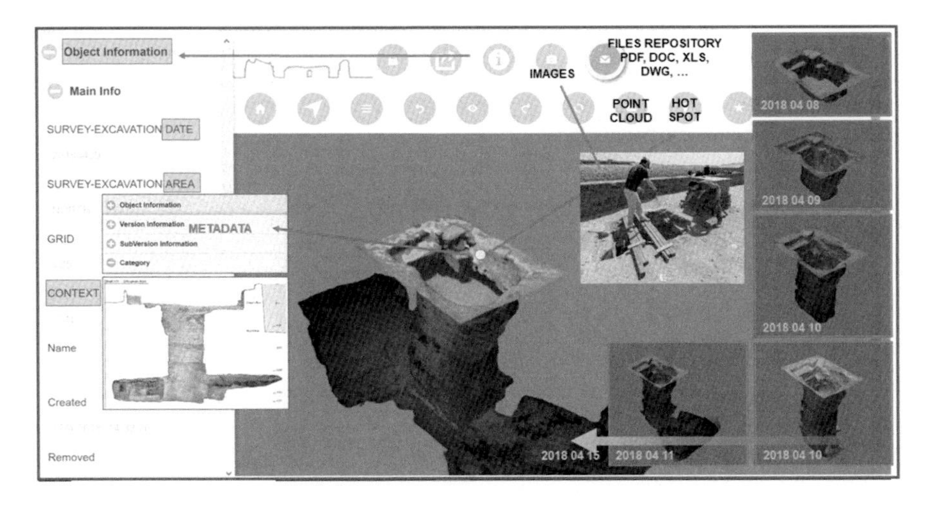

Fig. 3 The online system is expressly designed for the management, sharing and use of high-resolution 3D models and information following the excavation phases. The system allows measurement operations to be performed, along with the possibility to visualize very high-resolution orthophotos, place hotspots, linked documents and images

mode. The database is created to offer the possibility to add extra information systems created ad hoc for every single study case. This is the key aspect of the system that can be therefore easily adapted to completely different study cases.

In 2015, BIM3DSG became a component of the ERC project LIFE, for the management and visualization of the data collected during the archaeological expeditions to Umm al-Dabadib, in Egypt's Western Desert (Fassi et al. 2015). The initial work performed on these archaeological remains proved extremely useful and interesting, and prompted the development of a new branch of research, focusing on the construction of a version of the core system which would be specifically designed to respond to the needs of archaeological excavations (Fig. 3).

The main issue relating to the creation of a version of the system specifically designed for archaeology is the need to link time, space, objects and information, in order to visualize the past situation and allow for a comparison with the current situation or with the different phases of modification. An archaeological excavation is an ongoing process, during which items of various types are found and removed: the informative system accompanying the excavation must therefore be able to record in real time the physical transformation of the area under excavation and geo-reference both in space and time the findings that are progressively retrieved. Recording in 3D the excavation and its findings has a number of advantages but may always be feasible, due a combination of logistic, environmental and financial reasons. Whether or not 3D models of items and contexts are adopted, it is certainly necessary to attach to the finds different types of information, ranging from images to written notes. The ideal solution is to be able to do so from the smallest context to the landscape scale, thus connecting any findings to the wider framework of the site, as well as to the bigger picture.

Finally, considering that most of the specialized analysis on the findings is carried out at a later stage, and by researchers often scattered across various countries, the informative system is designed to offer a collaborative work environment, so that the data is available to the research team independently from their physical location.

6 Conclusions and Directions of Future Research

The spread of 3D models and the advantages of digitally recording data and information has an inevitable impact on archaeology, overlapping, missing out certain aspects and in some cases clashing with the traditional archaeological practice. The work carried out by the research team of the ERC project LIFE at the ABC Department focuses on solving these issues at various levels: by testing the most effective survey strategies, as well as by developing an informative system based on connectivity among data, information, places and people. The core of the system is being constantly updated and upgraded by the creation of new components and the fine-tuning of others, in strict collaboration with the archaeologists, who will be the final users of this product.

References

Adami A, Fassi F, Fregonese L, Piana M (2018) Image-based techniques for the survey of mosaics in the St Mark's Basilica in Venice. Virtual Archaeol Rev 9(19):1–20. https://doi.org/10.4995/var.2018.9087

Betrò M (2011) Virtual environments and web community in archaeology: Theban tomb 14 as case study. In: Belova GA (ed) Achievements and problems of modern egyptology. Proceedings of the international conference held in Moscow on Sept 29–Oct 2, 2009. Russian Academy of Sciences, Center for Egyptological Studies, pp 38–46

Del Vesco P (2018) 'Tutto ciò che ha valore è senza difese'. Archeologia e distruzioni. In AAVV, Anche le statue muoiono: conflitto e patrimonio tra antico e contemporaneo (catalogue of the exhibition Statues also die at Museo Egizio, Fondazione Sandretto Re Rebaudengo, Musei Reali Torino), Modena, pp 40–51

Del Vesco P, Greco C, Müller M, Staring N, Weiss L (2019) Current research of the Leiden-Turin archaeological mission in Saqqara. A preliminary report on the 2018 season. Rivista del Museo Egizio 3:1–25. https://doi.org/10.29353/rime.2019.2236

Fassi F, Parri S (2012) Complex architecture in 3D: from survey to web. Int J Heritage Digital Era 1:379–398

Fassi F, Achille C, Fregonese L (2011) Surveying and modelling the Main Spire of Milan Cathedral using multiple data sources. Photogram Rec 26:462–487

Fassi F, Rechichi F, Parri S (2014) Metodo e sistema per la gestione e la visualizzazione di modelli di oggetti tridimensionali complessi. In: Italian patent pending MI2014A002016

Fassi F, Rossi C, Mandelli A (2015) Emergency survey of endangered or logistically complex archaeological sites. In: International archives of the photogrammetry, remote sensing and spatial information sciences, vol XL-5/W4, pp 85–91. https://doi.org/10.5194/isprsarchives-xl-5-w4-85-2015

Fassi F, Fregonese L, Adami A, Rechichi F (2017) Bim system for the conservation and preservation of the mosaics of San Marco in Venice. In: International archives of the photogrammetry, remote sensing and spatial information sciences, vol XLII-2/W5, pp 229–236, https://doi.org/10.5194/isprs-archives-XLII-2-W5-229-2017

Fregonese L, Fassi F, Achille C, Adami C, Ackermann S, Nobile A, Giampaola D, Carsana V (2016) 3D survey technologies: investigations on accuracy and usability in archaeology. The case study of the new "Municipio" underground station in Naples. ACTA IMEKO 5(2):55–63

Greco C (2018) Il museo e la sua natura. AAVV, Anche le statue muoiono: conflitto e patrimonio tra antico e contemporaneo (catalogue of the exhibition Statues also die at Museo Egizio, Fondazione Sandretto Re Rebaudengo, Musei Reali Torino), Modena, pp 21–27

Mandelli A, Achille C, Tommasi C, Fassi F (2017) Integration of 3d models and diagnostic analyses through a conservation-oriented information system. In: International archives of the photogrammetry, remote sensing and spatial information sciences, vol XLII-2/W5, pp 497–504. https://doi.org/10.5194/isprs-archives-XLII-2-W5-497-2017

Martin GT (1991) The hidden tombs of Memphis. Egypt Exploration Society, London

Rechichi F, Mandelli A, Achille C, Fassi F (2016) Sharing high-resolution models and information on web: the web module of bim3dsg system. In: International archives of the photogrammetry, remote sensing and spatial information sciences, vol XLI-B5, pp 703–710. https://doi.org/10.5194/isprs-archives-XLI-B5-703-2016

Rossi C, Fiorillo F (2018) A metrological study of the Late Roman Fort of Umm al-Dabadib, Kharga Oasis (Egypt). Nexus Netw J 20(2):373–391

Tommasi C, Fiorillo F, Jiménez Fernández-Palacios B, Achille C (2019) Access and web-sharing of 3d digital documentation of environmental and architectural heritage. In: International archives of the photogrammetry, remote sensing and spatial information sciences, vol XLII-2/W9, pp 707–714. https://doi.org/10.5194/isprs-archives-XLII-2-W9-707-2019

Verhoeven G, Taelman D, Vermeulen F (2012) Computer vision-based orthophoto mapping of complex archaeological sites: the ancient quarry of Pitaranha (Portugal–Spain). Archaeometry 54:1114–1129. https://doi.org/10.1111/j.1475-4754.2012.00667.x

Tools for Reading and Designing the 'Islamic City'. Italian Urban Studies at the Crossroads

Michele Caja, Martina Landsberger and Cecilia Fumagalli

Abstract Starting from the typological and morphological studies conducted in Italy since the Sixties of the last century, the essay aims to investigate limits and possibilities in the application of these studies to urban contexts that are different from those for which they were originally conceived and developed. The rediscovery of the historic city, mostly forgotten or cancelled since then, both in the intentions and physically, receives from here an international acknowledgement. The researches on the Islamic urban phenomenon carried out by Italian scholars were strictly linked to design issues. Since the end of the 1960s, in fact, several Italian architects were involved in town planning activities of several cities of the Islamic world.

Keywords Typology · Morphology · Urban studies · Historic city · Islamic world

Starting from the beginning of the 1950s, when the urban principles defined and established by the Modern Movement started to be revised and discussed, a new research field comes to the fore: the European historic city starts to be a new area of interest, also thanks to the debate on the 'heart of the city' faced at 1951 CIAM in Hoddeshon (Tyrwhitt et al. 1952).

At the end of the 1950s, some Italian architects and researchers started their investigations on specific historic cities: Saverio Muratori and Paolo Maretto studied Venice and Rome (Muratori 1960; Maretto 1960; Muratori et al. 1963), Gianfranco Caniggia focused on Como (Caniggia 1963), Aldo Rossi dealt with Milan (Rossi 1964), Carlo Aymonino chose Padova (Aymonino et al. 1970) and Giorgio Grassi (Grassi 1971) and Antonio Monestiroli (Monestiroli 1973) worked on Pavia. These researches introduce a new way to look at the urban form, a method to understand and describe it, based on its typo-morphological characters.

M. Caja (✉) · M. Landsberger
Architecture, Built Environment and Construction Engineering—ABC Department, Politecnico di Milano, Milan, Italy
e-mail: michele.caja@polimi.it

C. Fumagalli
Ecole D'Architecture, Université Internationale de Rabat, Rabat, Morocco

N. Aste et al. (eds.), *Innovative Models for Sustainable Development in Emerging African Countries*, Research for Development, https://doi.org/10.1007/978-3-030-33323-2_14

Consequently, the rediscovery of the historic city, mostly forgotten or cancelled since then, both in the intentions and physically, receives an international acknowledgement.

Even though the Italian studies represent important contributions in the field, it is abroad that a more direct and strict connection between the analytical and the practical phases of the project has been investigated at most.

We can place in this framework, and according to the need for the reconstruction of the European historic cities, the researches developed by Jean Castex and Philippe Panerai on Versailles (Castex et al. 1980) and by Leon Krier and Maurice Culot on Brussels (Krier and Culot 1982), among others.

In order to understand the structure of the Islamic city, which seems to have been neglected by this group of scholars as a testing ground both for the design theories and the analytical studies, we should or could refer to the typo-morphological researches established in the framework of the European urban culture of the last century (Fig. 1).

Saverio Muratori (1910–1973) is one of the first Italian architects of the post-war period to adopt the notion of 'typology' in connection with the one of 'morphology': we can easily affirm that he has inaugurated the era of the applied urban analysis that have had a great influence and echo both nationally and internationally.

The first result of his researches is the book *Studi per una operante storia urbana di Venezia*, published in 1960, that collects all the field surveys and the analyses carried out by Saverio Muratori and his students at the IUAV of Venice.

The book is an articulate and complex work, able to render, with drawings, images and texts, the stratification in time and the historical dimension of the Venice urban fabric, through general location plans, drawings of the ground floors of public buildings and residential blocks, photographic surveys on specific neighbourhoods or parts of the town, and investigations on specific buildings according to different epochs.

The originality and the success of this book lie in the fact that it has been able to rediscover the richness and the complexity of the historic city in a period in which research in architecture was facing a deep crisis: the re-foundation ideals proposed by the Modern Movement were in fact being abandoned both because of the negative results of the reconstruction projects of the second post-war period and their inability of 'making city', especially if confronted to the consolidated and compact urban fabric of the historic city, of which Venice represents a valuable example (Fig. 2).

Thanks to studies similar to the one proposed by Muratori in Venice, the urban analysis rediscovers the morphological dimension of the residential fabric, considering the buildings not as single objects, but as elements of the compact and consolidated morphological structure, defined by different scales: the block, intended as portion of land delimited by streets and canals; the structure of the parcels showing the complex issue of property in the making of history; the building in its typo-morphological essence.

The use of ground floor plans of specific portions of Venice's urban fabric (which, apparently, seems to be simple and easy, but that is indeed the result of a complex entanglement between archival researches, cadastral investigations, on field verifications and graphic elaborations of the documentation at the same scale) highlights the strict relationship existing between public space, built environment and water,

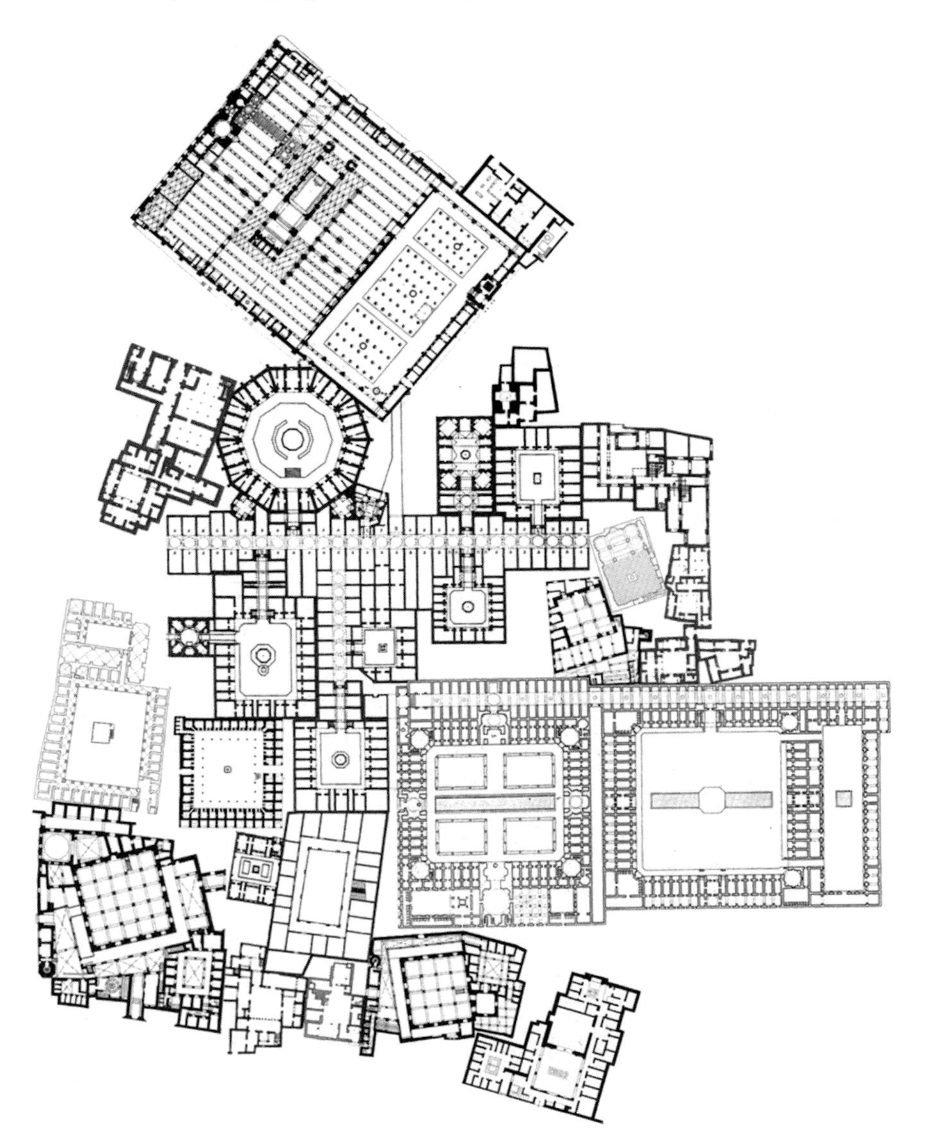

Fig. 1 Collage Islamic city (drawing by Cecilia Fumagalli)

Fig. 2 S. Muratori, Venezia: Urban Districts (Isola di San Lio), 1960, from S. Muratori, *Studi per una operante storia urbana di Venezia*, Istituto Poligrafico dello Stato, Roma, 1960, pag. 53, Tav. 4

able to represent the structure of the *calli, campi* and canals and the more private and domestic space of the entrance halls and inner courtyards.

Based on the strict morphological links within the investigated urban fabrics, it is possible to deduce the typological essence of the buildings.

The urban scale considered by Muratori in his researches integrates the more architecturally oriented scale adopted by Paolo Maretto, who also conducted a careful investigation on Venice in relation to gothic buildings.

In his book *L'edilizia gotica veneziana,* Maretto focuses on the transformations occurred on single buildings throughout history, thanks to accurate plans and rich archival documentation.

A similar analytical approach is the one adopted by Gianfranco Caniggia—one of Muratori's pupils—who chose Como as his testing ground. Thanks to a synchronic reading, Caniggia superimposes to the structure of the existing city the modular grid of the Roman settlement. This allows the Italian scholar to highlight different episodes within the urban fabric, from the *tipo base* defined as a common matrix, to the consequent *tipi differenziati.*

In this case, the type is intended as the institutional ensemble of building able to reproduce an organism as a synthetic fact, and at the same time, it possesses characters such as to allow the formation of an urban fabric adjacent to neighbouring buildings and in adherence with the road layout.

Exported abroad, especially to England, France, Belgium and Germany from the 1970s onwards (Merlin et al. 1988; Panerai et al. 1999), the continuity of such a typo-morphological approach applied to the investigations on the city and its architecture is still visible today in different school of thoughts that have focused their researches on the architectural organism and their attention on previous studies, such as those carried out by Conzen on Alnwick (Conzen 1960) or on some of the themes addressed by the International Conferences of Urban Form (ISUF), from 1994 onwards (Strappa 1995; Strappa et al. 2015; Caja et al. 2012, 2016).

Among the previously mentioned studies and researches, we should comprise also the less known works on the urban fabric of the Islamic city.[1] In fact, the Italian contribution on urban studies only apparently neglected the Islamic world. Even though it is not very well known and somehow relegated to the background by the official historiography, the Italian studies have had a great influence in the definition of the urban phenomenon in the Islamic world and in the formation of an overall knowledge about this issue.

In 1981, the Italian scholar and architect Paolo Portoghesi was appointed curator of the 2nd Venice Biennale devoted to the 'Architecture in Islamic Countries' (Cuneo et al. 1982).

The exhibition ratified the international acknowledgement of the themes and issues linked to the architecture and the city of the Islamic world, legitimizing the studies, the researches and the designs carried out since then by framing them in a wider scientific debate.

Several sections of the exhibition were devoted to the presentation of achieved and ongoing projects by well-known international architects such as Hassan Fathy, Louis Kahn, Le Corbusier, Fernand Pouillon, Kenzo Tange, SOM and by some Italian architects such as Vittorio Gregotti and the BBPR group.

Paolo Cuneo curated a section on the restoration and revitalization projects of the historic heritage of several cities, by framing the issue from a scientific point of view. The design dimension, which is the main focus of the exhibition and the reason for it, is scientifically framed by Ludovico Quaroni, who has the difficult task to investigate and present the origins of the urban Islamic phenomenon.

As for the cases quoted in the opening of the paper, the researches on the Islamic urban phenomenon carried out by Italian scholars were strictly linked to design issues.

Since the end of the 1960s, in fact, several Italian architects were involved in town planning activities of several cities of the Islamic world. It is, for example, the case of the Italian involvement in Tunisia, where, along with the issues posed by urban and territorial planning for the development and the modernization of the main cities, the Italian architects involved in the design activities, started to be more and more involved in the issues related to the historic centres. From 1961 to 1970, Ludovico

[1] Among the scientific community, the definition of 'Islamic city' is a still debated issue, and we are far from agreeing about a univocal one. For the purposes of the present paper, we suggest to consider the adjective 'Islamic' as a label able to represent a wide geographical and cultural frame to the discourse on the urban issue.

Quaroni worked at the *Plan Directeur du Grand Tunis*,[2] suggesting, among other things, the integration of the *medina* within the general urban strategies.

Some of the young architects working in Quaroni's design team started to focus their attention and their research interests towards the historic city of Tunis and started to be officially involved in the activities proposed by the public administrations concerning the realization of urban and architectural surveys of Tunis old city.

It is in those years that Roberto Berardi, one of the young architects of Quaroni's team, started his collaboration with the newly founded *Association pour la Sauvegarde de la Médina de Tunis*. In this framework, Berardi carried out a detailed analysis of the urban fabric, of the residential typologies and of the main collective buildings of the historic city of Tunis, that allowed him to establish a veritable urban analysis method that he applied also to other cities of the Islamic world (Privitera and Metalsi 2016; Berardi 1979, 2005, 2008).

Morphology is the instrument, the lens that Berardi adopted to read the city of Tunis and to exploit its analysis; composition is the tool, the operation that enabled him to understand its structure.

Moreover, the representation of the city through its ground plans, as it has been performed by Berardi, is able to render the most meaningful and precise idea of the city itself: 'it is, in some way, the design of that which preceded the stereometric of the city, and also the design of that which would remain of the city, if it were reduced to ruins. And it is also a palimpsest of transformations that have succeeded each other, overlying earlier phases' (Berardi 1989) (Figs. 3 and 4).

A similar experimentation of operative research on the urban fabric of an Islamic city is the one carried out in Algeria between the 1970s and 1980s by another group of architects and scholars in the framework of bilateral agreements for scientific and technical cooperation among Algeria and Italy. It is in indeed in Algiers, and thanks within the COMEDOR,[3] that a team composed by three Italian architects (Daniele Pini, Marcello Balbo and Corrado Baldi) and an Algerian sociologist (Sidi Boumediène) founded the *Atelier Casbah*, a research centre on the city and an incubator of ideas and studies for the rehabilitation of Algiers historic centre. The team conducted a meaningful research, presented in a monographic issue of the Italian magazine *Parametro* (Balbo et al. 1973), directed towards the morphological and typological reading of the residential urban fabric of the Casbah. If the Casbah group was involved in an operative research directed towards the drafting of a rehabilitation project for the historic city, the scientific cooperation established among Italian researchers from the Architecture Faculty of the University of Rome and the Ecole Polytechnique d'Architecture et d'Urbanisme of Algiers (EPAU) opens the way to the establishment of systematic studies on the Islamic city with the aim of defining its

[2]The project was carried out by the Bureau d'Etudes Ludovico Quaroni-Adolfo De Carlo, which comprised Ludovico Quaroni, Adolfo De Carlo, Massimo Amodei, Roberto Berardi and Benjamin Hagler.

[3]The COMEDOR, acronyme for *Comité Permanent d'Etudes, d'Organisation et de Developpement de l'Agglomeration d'Alger*, was the first Algerian public institution to be charged of the development of urban development strategies.

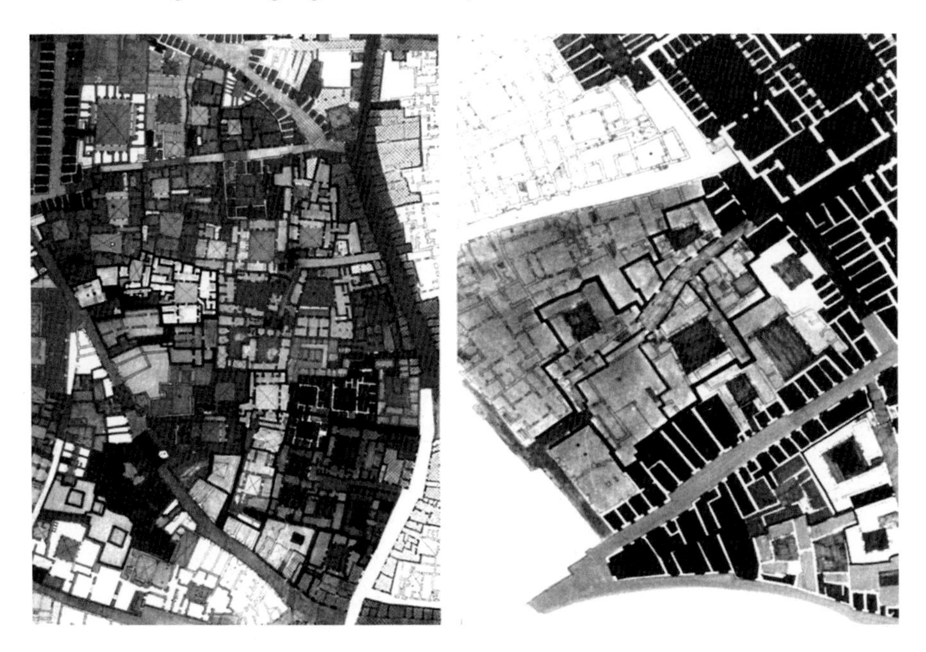

Fig. 3 Left: plan of Tunis Medina by Roberto Berardi, showing his morphological study. Right: fragment of the ground floors plan of Tunis Medina by Roberto Berardi (*source* Privitera and Metalsi 2016)

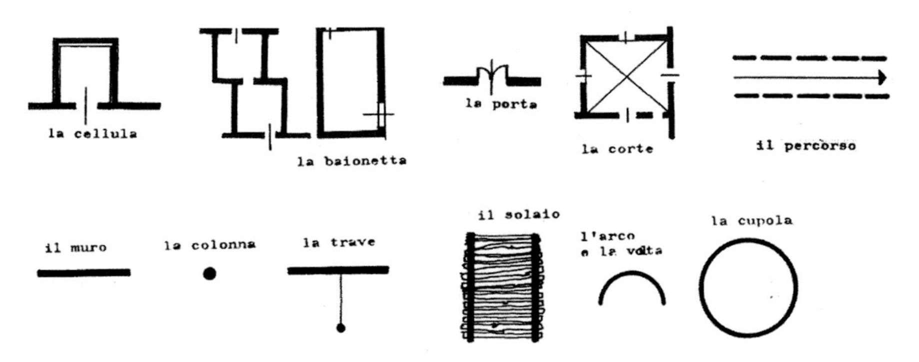

Fig. 4 The basic elements composing the urban fabric of Tunis according to Roberto Berardi (*source* Privitera and Metalsi 2016)

form and its structure. The different academic and scientific exchanges in Algeria represent, in fact, the occasion for some Italian scholars, such as Paolo Cuneo, Ludovico Micara and Attilio Petruccioli, to carry out a meaningful study on the urban phenomenon in the Islamic world. After these experiences, the three scholars published three different books presenting their researches and their respective points of view. In his *Storia dell'urbanistica. Il mondo islamico* (Cuneo 1986), Paolo Cuneo faced the issue of the definition of the character and the development of the cities of the

Islamic world according to a rigid subdivision into geographical areas and following an historic path until modern times. If Cuneo's encyclopaedic work was informed by a historic approach, on the other side, Ludovico Micara and Attilio Petruccioli followed the teachings of their masters Ludovico Quaroni and Saverio Muratori and adopted morphological and typological readings. Ludovico Micara (Micara 1985) chose to read the Islamic city through its collective institutions, highlighting typological variations and morphological issues. Attilio Petruccioli (Petruccioli 1985), convinced that the urban fabric of a city contained its history, suggested an analytical method aimed at deciphering the transformations occurred within the urban fabric through its structural reading. Moreover, the research approach suggested by the architect follows the study and the understanding of the design tools adopted to build the urban and territorial environment. If we want to sum up the themes discussed all along the paper, one element comes out clearly: the strict, symbiotic relationship between design and typo-morphological analysis, as if to say that one is pre-requisite or, on the contrary, the reason for the other. In other words, all the studies considered by the present paper show the necessity of typo-morphological considerations, in order to carry out any design activity, which intrinsically is a transformation of a previously established condition.

References

Aymonino C, Brusatin M, Fabbri G, Lena M, Lovero P, Lucianetti S, Rossi A (1970) La città di Padova. Saggio di analisi urbana. Officina, Roma

Balbo M, Moretti G, Baldi C, Sartori S, Dehò L, Pini D (1973) Parametro, 17

Berardi R (1979) Espace et ville en pays d'Islam. In: Chevallier D (ed) L'espace sociale de la ville arabe. Maisonneuve et Larose, Paris

Berardi R (2005) Saggi su città arabe del Mediterraneo sud orientale. Alinea, Firenze

Berardi R (1989) On the city. Environmental Design: Journal of the Islamic Environmental Design Research Centre 1–2. Carucci Editore, Roma, 11

Berardi R (2008) The spatial organization of Tunis medina and other Arab-Muslim cities in North Africa and the Near East. In: Salma KJ, Holod R, Petruccioli A, Raymond A (eds) The city in the Islamic world, vol 1. Brill, Leiden, Boston, pp 269–283

Caja M, Landsberger M, Malcovati S (eds) (2012) Tipologia architettonica e morfologia urbana. Il dibattito italiano – antologia 1960–1980. Libraccio, Milano

Caja M, Landsberger M, Malcovati S (eds) (2016) Tipo forma figura. Il dibattito internazionale 1970–1990. Libraccio, Milano

Caniggia G (1963) Lettura di una città: Como. Centro studi di Storia Urbanistica, Roma

Castex J, Céleste P, Panerai P (1980) Lecture d'une ville: Versailles. Editions du Moniteur, Paris

Conzen MRG (1960) Alnwick, Northumberland: A study in town-plan analysis. Inst Br Geogr 27:3–122

Cuneo P, Micara L, Petruccioli A (eds) (1982) Architettura nei Paesi Islamici. Seconda mostra internazionale di architettura. Edizioni La Biennale di Venezia, Venezia

Cuneo P (1986) Storia dell'urbanistica. Il mondo islamico. Laterza, Roma-Bari

Grassi G (1971) Tipologie d'abitazione a Pavia. Controspazio 9

Krier L, Culot M (1982) Contreprojets. In: A.A.M (ed) La reconstruction de Bruxelles. A.A.M, Bruxelles

Maretto P (1960) L'edilizia gotica veneziana. Istituto Poligrafico dello Stato, Roma

Merlin P, D'Alfonso E, Choay F (eds) (1988) Morphologie urbaine et parcellaire. Presses Universitaires de Vincennes, Saint-Denis
Micara M (1985) Architetture e spazi dell'Islam. Le istituzioni collettive e la vita urbana. Carucci editore, Roma
Monestiroli A (1973) Pavia: storia e progettazione della città. Edilizia Popolare 113
Muratori S (1960) Studi per una operante storia urbana di Venezia. Istituto Poligrafico dello Stato, Roma
Muratori S et al (1963) Studi per una operante storia urbana di Roma. Consiglio Nazionale delle Ricerche, Roma
Panerai P, Depaule J-C, Demorgon M (1999) Analyse urbaine. Parenthèse, Marseille
Privitera F, Metalsi M (2016) Les signes de la Médina. La morphologie urbaine selon Roberto Berardi. Dipartimento di Architettura di Firenze, Firenze
Petruccioli A (1985) Dar-al-Islam: architetture del territorio nei paesi islamici. Carucci editore, Roma
Rossi A (1964) Contributo al problema dei rapporti tra tipologia edilizia e morfologia urbana. Esame di un'area di studio di Milano con particolare attenzione alle tipologie edilizi prodotte da interventi privati. ILSES, Milano
Strappa G (1995) Unità dell'organismo architettonico, Note sulla formazione e trasformazione dei caratteri degli edifici. Dedalo, Bari
Strappa G, Amato ARD, Camporeale A (eds) (2015) City as organism: new visions for urban life. ISUF, Rome
Tyrwhitt J, Sert JL, Rogers EN (eds) (1952) The heart of the city. Pellegrini and Cudah, New York